Pascal's Arithmetical Triangle

By the same author

Likelihood
Cambridge University Press, 1972

Foundations of Mathematical Genetics
Cambridge University Press, 1977

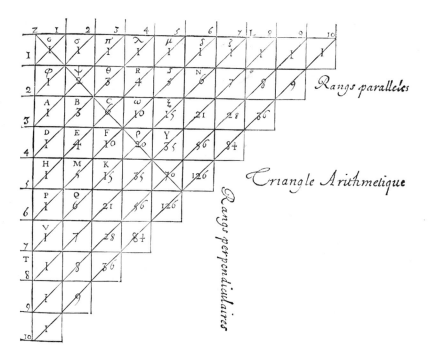

Pascal's original Arithmetical Triangle, from the *Traité du triangle arithmétique* (1665)

Pascal's

Arithmetical Triangle

A.W.F. EDWARDS, Sc.D.

Reader in Mathematical Biology, University of Cambridge
Fellow of Gonville and Caius College

512.7
E

CHARLES GRIFFIN & COMPANY LTD
London
OXFORD UNIVERSITY PRESS
New York

CHARLES GRIFFIN & COMPANY LIMITED
16 Pembridge Road, London W11 3HL, U.K.

First published 1987

Edwards, A. W. F. (Anthony William Fairbank), 1935—
    Pascal's arithmetical triangle.

    Bibliography: p.
    Includes index.
    1. Pascal's triangle.   I. Title.
QA241.E37   1986   512'.57   86-19264
ISBN 0-19-520546-4

Edwards, A.W.F.
    Pascal's arithmetical triangle.
    1. Probabilities——History
    I. Title
    519.2'09   QA273.A4

    ISBN 0-85264-283-0

Typeset in Great Britain by
Kingprint Limited, Richmond, Surrey

Printed & bound in Great Britain by
Redwood Burn Limited, Trowbridge, Wilts

For CATHARINA

PASCAL'S ARITHMETICAL TRIANGLE

Inexhaustae Scientiae TABVLA secretissima Arithmetices Arcana pandens.

FAULHABER (attrib.), 1615

[A rich mine of information in the form of a TABLE which unfolds the most recondite secrets of arithmetic.]

Plena haec mysteriis pulcherrimis tabella.   OUGHTRED, 1631

[This table full of the most beautiful mysteries.]

Abacus ΠΑΓΧΡΗΣΤΟΣ.   BRIGGS, 1633

[A calculator useful for everything.]

C'est une chose étrange combien il est fertile en propriétés.   PASCAL, 1654

Habet haec Tabula proprietates plane eximias & admirandas; praeterquam enim quod Combinationum mysterium in illa latere jam ostendimus, notum est interioris Geometriae peritis, praecipua etiam totius reliquae Matheseos arcana inibi delitescere.   JAMES BERNOULLI, 1713

[This Table has truly exceptional and admirable properties; for besides concealing within itself the mysteries of Combinations, as we have seen, it is known by those expert in the higher parts of Mathematics also to hold the foremost secrets of the whole of the rest of the subject.]

The contemplation of this table has probably been attended with the invention and extension of some of our most curious discoveries in mathematics, both in regard to the powers of a binomial, with the consequent extraction of roots, the doctrine of angular sections by Vieta, and the differential method by Briggs and others.

HUTTON, 1786

Contents

NOTE

All works cited thus: Pascal (1665), in the notes at the end of each chapter, are listed in the *References* at end of book. The date used is that of the original publication of the work cited, though the page reference may be to a more recent edition. Works written before the ˙ .vention of printing are given the date of writing, which will often be only approximate. There are no references listed with dates Before Christ, so that all dates are to be interpreted A.D. In the case of references to material that was not published for many years after it was written, such as Fermat's correspondence, the date of writing is used.

The intention is thus to combine a reference system with an approximate chronology, though the great variety of source material has inevitably led to many compromises.

Preface

The Arithmetical Triangle is the most famous of all number patterns. Apparently a simple listing of the binomial coefficients, it contains the triangular and pyramidal numbers of ancient Greece, the combinatorial numbers which arose in the Hindu studies of arrangements and selections, and (barely concealed) the Fibonacci numbers from medieval Italy. It reveals patterns which delight the eye, raises questions which tax the number-theorists, and amongst the coefficients "There are so many relations present that when someone finds a new identity, there aren't many people who get excited about it any more, except the discoverer!" [1].

It was first written down long before 1654, the year in which Blaise Pascal wrote his *Traité du triangle arithmétique*, but it was this work that brought together all the different aspects of the numbers for the first time. In it Pascal developed the properties of the numbers as a piece of pure mathematics (often using mathematical induction in his proofs) and then, in a series of appendices, showed how these properties were relevant to the study of the figurate numbers, to the theory of combinations, to the expansion of binomial expressions, and to the solution of an important problem in the theory of probability. That the Arithmetical Triangle should bear Pascal's name cannot be disputed.

The subsequent flowering of the theory of probability gave added importance to the binomial coefficients through the central role played by the binomial distribution, and in the development of analysis in the seventeenth century they cropped up at every turn, directly influencing the discoveries of Wallis, Newton and Leibniz. More recently the growing importance of combinatorial theory has fuelled fresh interest in the coefficients.

In the history of these numbers it will be important to differentiate between the three principal ways in which they arise. We shall therefore distinguish three types of numbers. By the *figurate numbers* we shall mean the numbers which occur in a certain family of progressions in which the lth number of the kth progression, denoted by f_k^l, is the sum of the first l numbers in the previous, or $(k-1)$th, progression, the first such progression being the series of integers, $1, 2, 3, 4, \ldots$; by the *combinatorial numbers* we shall mean the numbers which give the number of combinations of n different things taken r at a time, denoted by ^nC_r; and by the *binomial numbers* we shall mean the coefficients which occur in the expansion of the nth power of the binomial expression $(a+b)$, the coefficient of $a^r b^{n-r}$ being denoted by $\binom{n}{r}$. Analogously, an array of numbers will be called a *Figurate Triangle*, a *Combinatorial Triangle*, or a *Binomial Triangle*, according to the context in which it has been given. Such arrays will not necessarily be triangular—indeed, since they are capable of indefinite extension they can hardly be thought of as having any particular shape—but it is convenient, and consistent, to use Pascal's word. Of course, all these Triangles turn out to be the same thing, and once we have established their identity we will be able to relax this artificial terminology and refer to *Pascal's Arithmetical Triangle*, and to its numbers as simply *binomial coefficients*, the dominant name for them.

It was Montmort [2] who, in 1708, first attached the name of Pascal to the Combinatorial Triangle ("*Table de M. Pascal pour les combinaisons*"), though his array was not actually in Pascalian format (see Frontispiece) but a staggered version of it (Fig. 1).

$$1 . 1 . 1 . 1 . 1 \ldots$$
$$1 . 2 . 3 . 4 \ldots$$
$$1 . 3 . 6 \ldots$$
$$1 . 4 \ldots$$
$$1 \ldots$$

Fig. 1   Montmort's arrangement of the Combinatorial Triangle (1708)

Then, in his *Miscellanea Analytica* of 1730, De Moivre christened Pascal's form "*Triangulum Arithmeticum PASCALIANUM*" [3], reserving the plain "*TRIANGULUM ARITHMETICUM*" [4] for Montmort's format. It is to De Moivre, therefore, that I owe the title of this book.

There appears to be no previous full-length history of the Arithmetical Triangle, and in English not even a brief one. For German readers there are Tropfke [5] (1924) and Knobloch [6]

(1973) and for French readers the somewhat inaccessible account of Bosmans [7] published in 1906. Of the many extant tables of the Arithmetical Triangle we may note especially the beautiful *Table of Binomial Coefficients* edited by Miller [8] just before computers rendered such compilations obsolete.

Although the present work does not set out to provide a comprehensive account of Pascal's contribution to the theory of probability, it covers the greater part of it. To render it complete in this respect two of the author's papers, describing the Problem of Points and the Gambler's Ruin Problem, are reprinted as appendices.

During the writing I have been deeply indebted to Dr D. T. Whiteside, who, when he might so justifiably have concluded from my first page of notes that my cause was hopeless, instead embarked on a long correspondence which has directed, corrected, encouraged, and reprimanded me in just the right proportions. It is small wonder that earlier writers in this field made so many mistakes, for they did not have his fund of knowledge to call upon. But he did not, alas, write this book, whose mistakes are all my own.

Next it is a pleasure to thank the Master and Fellows of Gonville and Caius College, not only in their corporate capacity for possessing original editions of Bernoulli, De Moivre, Hérigone, Montmort, Oughtred, Vieta and Wallis and for providing excellent working conditions in College, but also individually for answering a torrent of questions, mathematical, logical, linguistic, and historical. The following colleagues have borne the brunt of these: Dr W. J. Macpherson, Dr J. Needham, Professor C. O. Brink, Reverend J. V. M. Sturdy, Dr E. F. Timms, Dr J. E. J. Altham, Dr J. E. Conway, Dr K. J. Martin, Dr P. J. Bayley, Dr R. H. S. Carpenter, Dr T. J. Pedley and Dr A. D. Barbour. Miss Sarah Law, also of Caius College, kindly read the whole text and made many helpful suggestions, whilst Miss Ann Edwards, of Trinity Hall, similarly helped with the references.

Finally, I am grateful to many correspondents for their help, which I acknowledge in the notes accompanying the chapters.

A. W. F. EDWARDS

Gonville and Caius College, Cambridge

Notes to Preface

[1] Knuth (1973), p. 53.
[2] Montmort (1708), p. 80.
[3] De Moivre (1730), p. 181.
[4] De Moivre (1730), p. 140.
[5] Tropfke (1924), pp. 34–45.
[6] Knobloch (1973), pp. 1–23.
[7] Bosmans (1906).
[8] Miller (1954).

1

The figurate numbers

The longest of the threads which Pascal wove together in his *Treatise on the Arithmetical Triangle* concerns the figurate numbers, and stretches back to the Pythagorean preoccupation with number-patterns 540 years before Christ [1].

When we need a concrete representation of an integer n we think immediately of n objects, such as pebbles, all alike. We may think of a handful of pebbles, or a bagful, where the positions of the individual pebbles signify nothing, but more probably we will have in our mind's eye a picture of the pebbles spread out on a table, or on the ground, so that each can be seen and counted. Asked which is the greatest of two numbers we would arrange pebbles in paired rows, noting which row was longer, and from such arrangements in one dimension we would surely start to play with patterns in two dimensions. The Pythagoreans considered triangular and square patterns and discovered that the corresponding *classification* of numbers led immediately to a *theory* of numbers.

The successive triangular numbers (Fig. 2) are 1, $1 + 2 = 3$, $3 + 3 = 6$, $6 + 4 = 10, \ldots, f_2^{l-1} + l = f_2^l, \ldots$, where f_2^l stands for the lth (the subscript 2 will change later as we generalize the numbers to more than two dimensions) and $f_2^1 = 1$. Obviously not all numbers are triangular numbers, so we have the beginnings of a scheme of classification. The sum of the first l integers is evidently the triangular number f_2^l, by virtue of the fundamental relation

$$f_2^l = f_2^{l-1} + l. \tag{1.1}$$

Next come the square numbers (Fig. 3), the lth consisting of l rows each of l dots, and thus being equal to l^2. To obtain one square number from its predecessor, dots forming a *gnomon* must be added

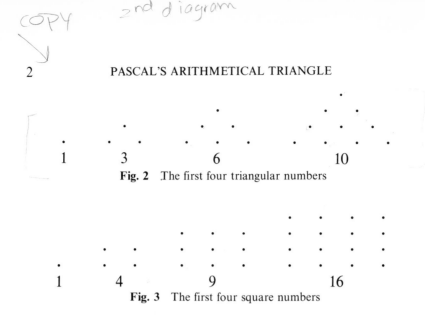

Fig. 2   The first four triangular numbers

Fig. 3   The first four square numbers

in the manner of Fig. 4, and from the fact that just $(2l + 1)$ dots are needed we learn that

$$(l + 1)^2 = l^2 + (2l + 1) \tag{1.2}$$

Fig. 4   A gnomon (here 7) added to a square number (9) makes the next square number (16)

and that, since the sequence $(2l + 1)$, $l = 0, 1, 2, 3, \ldots,$ is the sequence of odd numbers, the sum of the first l odd integers is a square. Moreover, as was known to Theon of Smyrna and to Nicomachus [2] at the beginning of the second century A.D., successive triangular numbers sum to a square (Fig. 5):

$$f_2^{l-1} + f_2^l = l^2. \tag{1.3}$$

Fig. 5   Successive triangular numbers (here 6 and 10) added together form a square number (16)

Eliminating f_2^{l-1} between (1.1) and (1.3), we find

$$f_2^l = \tfrac{1}{2}l(l+1),\tag{1.4}$$

whose geometrical interpretation is that twice the lth triangular number is an oblong with one more row than column (Fig. 6); but we should now regard it simply as the solution to (1.1) viewed as a recurrence equation.

Fig. 6   Twice a triangular number (here 6) is an oblong with one more row than column (4×3)

From square numbers we could proceed to other *polygonal* numbers in the plane, based on pentagons, hexagons, and so forth, but the generalization of triangular numbers which leads in the direction of the Arithmetical Triangle is not to square numbers in the plane but to triangular numbers in three dimensions, the *pyramidal* or *tetrahedral* numbers described by Theon and Nicomachus as being formed by summing the first l triangular numbers (Fig. 7), just as those numbers themselves had been formed by summing the first l natural numbers. Thus, extending our notation,

$$f_3^l = f_3^{l-1} + f_2^l, \; f_3^1 = 1.\tag{1.5}$$

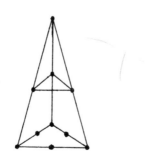

Fig. 7   The tetrahedral number ten formed from the first three triangular numbers

Nicomachus gave the tetrahedral numbers ("triangular pyramids" as opposed to pyramids on square or other bases) as

"1   4   10   20   35   56   84   and so on",

formed by the above rule from the triangular numbers

1   3   6   10   15   21   28   .   .   .

which are themselves formed from the integers

1   2   3   4   5   6   7   .   .   .

which are, of course, themselves formed by application of the same rule to the sequence

1   1   1   1   1   1   1   .   .   ..

But neither he nor Theon gave these numbers in tabular form (Fig. 8), nor did they apply the rule of formation to the tetrahedral numbers to produce another row, which would have corresponded to proceeding beyond three dimensions.

$$l$$

		1	2	3	4	5	6	7	.	.	.
	0	1	1	1	1	1	1	1	.	.	.
k	1	1	2	3	4	5	6	7	.	.	.
	2	1	3	6	10	15	21	28	.	.	.
	3	1	4	10	20	35	56	84	.	.	.

Fig. 8   Part of the Figurate Triangle: the figurate numbers in 0, 1, 2 and 3 dimensions

We can, of course, neither claim the Figurate Triangle for Pythagorean arithmetic, nor deny the possibility that it was written down, for the construction rules were known. In our notation the fundamental definition is:

$$f_k^l = f_k^{l-1} + f_{k-1}^l,$$

$$f_k^1 = f_0^l = f_0^1 = 1; \quad l = 2, 3, 4, \ldots; \quad k = 1, 2, 3, \ldots. \quad (1.6)$$

The alternative and equivalent form is:

$$f_k^l = \sum_{i=1}^{l} f_{k-1}^i, \quad (1.7)$$

as is easily proved,

There exists a papyrus from Ancient Egypt [3] (*ca* 300 B.C.) in which the scribe not only gave $\frac{1}{2}(l^2 + l)$ as the sum of the first l integers (i.e. equation 1.4) but continued with a statement equivalent to $\frac{1}{3}(l + 2)\frac{1}{2}(l^2 + l)$ for the sum of the first l triangular numbers. Gillings [4] comments "and if he proceeded to the next series, he would have written" the sum of the first l tetrahedral numbers as $\frac{1}{4}(l + 3)\frac{1}{3}(l + 2)\frac{1}{2}(l^2 + l)$, suggesting that the scribe anticipated the general rule for a figurate number (equation 1.18 below).

Be that as it may, the generalization of the figurate numbers to four and more dimensions did not occur in the West until over 1300 years after the time of Nicomachus and Theon, even though the shackles of dimension and geometry had been broken in other contexts by Diophantus of Alexandria in about A.D. 250. As we shall see in Chapter 4, Nicolo Tartaglia, an Italian mathematician, essentially extended Fig. 8 indefinitely in 1523 whilst enumerating dice throws, though his table was not published until 1556 [5].

Meanwhile in 1544 in Germany Stifel [6] gave a form of the extended Figurate Triangle in connection with the extraction of roots. The classical method for extracting a square root depends on the Euclidean identity $(a + b)^2 = a^2 + 2ab + b^2$ in the form $(a + b)^2 = a^2 + (2a + b)b$, a being the trial value; the method of setting out the working will have been learnt by some readers at school, but is not now so common.

The corresponding method for the cube root relies on $(a + b)^3 = a^3 + (3a^2 + 3ab + b^2)b$, and was known in the East before appearing in Italy in the work of Fibonacci in 1202 [7]. Stifel applied the method to roots of arbitrary degree, for which purpose he needed to find the binomial numbers of corresponding degree. However, he did not use the familiar additive procedure for deriving a Binomial Triangle (see Chapter 5), but proceeded first to derive a Figurate Triangle and then to obtain the binomial numbers from it. His method of construction was first to write down the series of natural numbers 1, 2, 3, 4, . . ., then next to the 3 to write down the series of triangular numbers (Stifel names them explicitly) 3, 6, 10, 15, . . ., then next to the 10 to write down the tetrahedral numbers 10, 20, 35, 56, . . ., found by using equation (1.6), and so on. In the resulting stepped triangle (Fig. 9) each step is two numbers deep, and each column is headed by the number to its left repeated [8]. By his method of construction we see that Stifel has appreciated that

$$f_m^m = f_{m-1}^{m+1}. \tag{1.8}$$

Der Ander theyl

1							
2							
3	3						
4	6						
5	10	10					
6	15	20					
7	21	35	35				
8	28	56	70				
9	36	84	126	126			
10	45	120	210	252			
11	55	156	330	462	462		
12	66	220	495	792	924		
13	78	286	715	1287	1716	1716	
14	91	364	1001	2002	3003	3432	
15	105	455	1365	3003	5005	6435	6435
16	120	560	1820	4368	8008	11440	12870

Es kan aber ein fleissiger Leser/diser tafel brauch leichtlich se=hen/aus den gesetzten satzungen der puncten/ Item auch wie sich die zalen der tafel aus einander finden / wer sich aber selbs nicht kan drauß verrichten/mag jm solliche zeygen lassen/ wie ich denn gnugsam dauon geschriben hab in meiner Latinischen Arithme=tica.

Fig. 9   Stifel's version of the Figurate Triangle (1545)

Then, in using his table to find the binomial numbers which he needs in the extraction of roots, Stifel essentially extends it by invoking symmetry about the last number in each even-numbered row, and about the last two numbers in each odd-numbered row. If the identity of the binomial numbers and the figurate numbers according to the equation

$$\binom{n}{r} = f_r^{n-r+1} \tag{1.9}$$

may be said to have had a discoverer in the West, he was surely Stifel.

Stifel's contemporary Scheubelius [9] was only a year after him in giving a Triangle in connection with the extraction of roots by the same method, but his Triangle is much more nearly Binomial in construction, so that we will describe it in Chapter 5 (though conscious of having made a very fine distinction). From this time on the identity of the figurate numbers and the binomial numbers was more and more widely appreciated; indeed the latter were often called "figurate". Even so, it was not always clear *why* the figurate numbers cropped up, and a good example of this is afforded by Vieta's discovery [10], in about 1591, of how to express $2\cos(\frac{1}{2}nx)$ and $2\sin(\frac{1}{2}nx)$ in terms of $2\cos(\frac{1}{2}x)$. Let $u_n = 2\cos(\frac{1}{2}nx)$; then Vieta proved [11] that

$$u_{n+1} = u_1 \cdot u_n - u_{n-1}, \quad n = 1, 2, 3, \ldots;$$
$$u_0 = 2; \quad u_1 = 2\cos(\frac{1}{2}x), \tag{1.10}$$

whence

$$u_0 = 2$$
$$u_1 = u_1$$
$$u_2 = u_1^2 - 2$$
$$u_3 = u_1^3 - 3u_1 \tag{1.11}$$
$$u_4 = u_1^4 - 4u_1^2 + 2$$
$$u_5 = u_1^5 - 5u_1^3 + 5u_1$$

$$\cdot \quad \cdot \quad \cdot \quad \cdot \quad \cdot \quad \cdot \quad \cdot$$

Omitting the first column and the negative signs (which occur in every alternate column), we have Vieta's table *Numeri multiplicium adfectionis* [12] for the coefficients:

$$
\begin{array}{llll}
2 & & & \\
3 & & & \\
4 & 2 & & \\
5 & 5 & & \\
6 & 9 & 2 & \\
7 & 14 & 7 & \\
8 & 20 & 16 & 2 \\
\end{array}
$$

. . . . .

each column of which is generated from the previous one in the usual manner of the figurate numbers, but "not starting with unity, as in the generation of powers, but starting with two" (Vieta treats the "generation of powers" elsewhere in his writings [13]). Thus each column consists of the first differences of the next.

For the sines, Vieta proved

$$v_{n+1} = u_1 \cdot v_n - v_{n-1}, \quad n = 1, 2, 3, \ldots,$$

$$v_0 = 0; \quad v_1 = 2 \sin(\tfrac{1}{2}x), \tag{1.13}$$

where $v_n = 2 \sin(\tfrac{1}{2}nx)$.

Then

$$v_1 = v_1$$

$$v_2 = v_1 u_1$$

$$v_3 = v_1(u_1^2 - 1)$$

$$v_4 = v_1(u_1^3 - 2u_1) \tag{1.14}$$

$$v_5 = v_1(u_1^4 - 3u_1^2 + 1)$$

. . . .

Omitting the first column and the signs, this table of coefficients is

$$
\begin{array}{llll}
1 & & & \\
2 & & & \\
3 & 1 & & \\
4 & 3 & & \\
5 & 6 & 1 & \\
6 & 10 & 4 & \\
7 & 15 & 10 & 1 \\
. & . & . & . \quad . \quad .,
\end{array}
\qquad (1.15)
$$

the columns now being the ordinary figurate numbers "as in the generation of powers", A century later the intimate connection between the series for cos nx and sin nx and the binomial numbers was clarified by the discovery of De Moivre's theorem

$$(\cos x + i \sin x)^n = \cos nx + i \sin nx. \qquad (1.16)$$

Vieta was the first person to name the higher figurate numbers; after "*numeri trianguli*" and "*pyramidales*" we have "*triangulo-trianguli*" and "*triangulo-pyramidales*". Later in this chapter we note Fermat's use of the names, but Fermat will have got them from Vieta, as will Oughtred [14], who explicitly pointed out the identity of the figurate numbers and the binomial numbers in his *Clavis Mathematicae* of 1631 [15]. Oughtred graduated from Cambridge in 1596 and became an influential private teacher of mathematics, including John Wallis amongst his pupils [16], whilst Newton possessed a copy of the third edition of *Clavis Mathematicae* [17]. These are important links, as we shall see in Chapter 8.

In the year of the publication of Oughtred's book, 1631, Henry Briggs, the inventor of common as opposed to Napierian logarithms, died, leaving in manuscript his *Trigonometria Britannica*, which was published by his friend Henry Gellibrand two years later [18]. Gellibrand says in a preface that Part I of Briggs's *Introduction* has been in manuscript for about thirty years; this *Introduction* contains a chapter [19] on the Figurate Triangle which briggs calls the "*Abacus ΠΑΓΧΡΗΣΤΟΣ*", or "calculator of many uses". Referring to the numbers as figurate numbers, Briggs points out their uses in binomial expansions and in trigonometrical calculations (following Vieta) and gives the addition rule of formation (1.6). But he then takes the matter one step further by pointing out (in words) that

$$kf_k^l = lf_{k-1}^{l+1}, \quad k = 2, 3, 4, \ldots: \tag{1.17}$$

"Any one of these diagonal numbers is in proportion to the next higher in the diagonal, as the *vertical* of the former is to the *marginal* of the latter". This is Hutton's translation [20] (our italics); *vertical* and *marginal* refer to the numbers at the heads of the columns and rows of the table (our l and $k + 1$, since Brigg's row numbers exceed ours in Fig. 8 by 1).

The repeated application of equation (1.17) enabled Briggs to find numbers in the Figurate Triangle without filling out the whole Triangle up to the diagonal in question. Such application is equivalent to:

$$f_k^l = \frac{l(l + 1)(l + 2) \ldots (l + k - 1)}{k(k - 1)(k - 2) \ldots 1}, \tag{1.18}$$

since $f_0^{l+k} = 1$ by (1.6), though Briggs also proceeded crab-wise across the table using a combination of the additive rule (1.6) and the multiplicative (1.17); substituting for f_k^{l-1} in the former by means of the latter leads directly to

$$f_k^l = \frac{l + k - 1}{k} f_{k-1}^l. \tag{1.19}$$

Although in the context of the present chapter the interest of (1.17), (1.18) and (1.19) is that they are relations amongst the figurate numbers known by Briggs from about 1600, we should note (in preparation for our discussion of the binomial expansion) that since Briggs explicitly mentioned raising a binomial expression to an integral power as one of the uses of his *Abacus*, equation (1.18) is the essence of the binomial theorem for positive integral index. Briggs, having been professor of geometry at Gresham College, London, from 1596 to 1619 (and thereafter professor in Oxford), may have had some familiarity with the great Italian Renaissance writers Cardano and Tartaglia and could have acquired his initial knowledge of the Arithmetical Triangle from them; but although Cardano gave equation (1.17) in words in 1570 (as we shall see) he did so in a combinatorial context, so that Briggs's is the first *explicit* Western linking of it with the binomial expansion [21].

We may also notice that there is, in Briggs's work, the beginnings of a theory of interpolation by finite differences [22]. In his *Arithmetica logarithmica* [23] of 1624 he not only carried out quadratic interpolation using second-order differences, but knew that the nth-order differences of the nth powers of the integers are constant. "Briggs . . . implicitly gives the basis for interpolation with regard to functions (tabulated at equal intervals of the argument) whose nth differences are constant–functions, that is (as Newton at least was to see), whose representing polynomial is of finite degree n", Whiteside [24] tells us, and "It is clear . . . that he had made a profound study of the nth-order finite differences. In later chapters of *Arithmetica logarithmica* he gives, without prior investigation or justification, rules which contain implicitly the general 'Newton–Gauss' interpolation formula" [25]. This formula and many related ones are of interest to a study of the Arithmetical Triangle because of the appearance of the figurate numbers in them, as had already been noted in the earliest Western treatise to consider finite differences, the remarkable *De Numeris Triangularibus et inde De Progressionibus Arithmeticis* of Thomas Harriot [26]. This manuscript, even now unpublished, was probably written in 1611 or a little earlier, and never seen by Newton [27] (or, very likely, by anyone else for that matter [28]).

In *De Numeris Triangularibus* Harriot demonstrates a knowledge of (1.18) as well as of the addition rule of formation of the figurate numbers, multiplying out (1.18) for values of k up to 7. Then, rearranging his table of figurate numbers (compare our Fig. 8) into a table of "binomial" numbers (Fig. 10), he notes that the successive rows sum to 1, 2, 4, 8, 16, . . . and that the number in the $(n + 1)$th row and $(r + 1)$th column is

$$\frac{n(n - 1)(n - 2)\ldots(n - r + 1)}{1 \cdot 2 \cdot 3 \ldots r}. \tag{1.20}$$

```
1   =   1
1   1   =   2
1   2   1   =   4
1   3   3   1   =   8
1   4   6   4   1   =   16
1   5   10  10  5   1   =   32
1   6   15  20  15  6   1   =   64
                              &c
```

Fig. 10   Harriot's second arrangement of the Figurate Triangle (1611)

Harriot indicates that his tables can be extended indefinitely, and even extended the table of figurate numbers backwards by subtraction (as if by using equation 1.6 in reverse to adjoin rows for $k = -1, -2, \ldots, -6$ to Fig. 8). He also possessed (1.19), and employed a double-index notation for the figurate numbers (as we have); he knew that (1.20) could accommodate negative or fractional values of n [29]. The transformation from (1.18) to (1.20) which accompanies the changed shape of the table was (we shall argue in Chapter 8) to be an important element in Newton's discovery of the general binomial theorem half a century later.

Harriot then gives what we now know as the Newton–Gregory forward-difference formula, which may be rendered in a modern notation as

$$f\left(\frac{k}{m}\right) = f(0) + \frac{k}{m}\Delta + \frac{\left(\frac{k}{m}\right)\left(\frac{k}{m}-1\right)}{2!}\Delta^2 + \frac{\left(\frac{k}{m}\right)\left(\frac{k}{m}-1\right)\left(\frac{k}{m}-2\right)}{3!}\Delta^3$$

$$+ \ldots, \quad (1.21)$$

in which $\Delta^r = \Delta^r f(0)$ is the rth leading unit difference (defined below), the series terminating with the term in Δ^n such that Δ^{n+1} and all higher-order differences are zero. The representation of the argument as a fraction k/m is to facilitate interpolation between unit values of the argument. Harriot knows, of course, that for integral k/m the coefficients of (1.21) are the figurate numbers in the form (1.20).

Consider a function $y = f(x)$ tabulated for integer arguments $x = 0, 1, 2, 3, \ldots$. Let $f(x + 1) - f(x)$ be defined as $\Delta f(x)$, the sequence of first differences, let $\Delta f(x + 1) - \Delta f(x) = \Delta^2 f(x)$, the second differences, and in general let

$$\Delta^{n-1} f(x + 1) - \Delta^{n-1} f(x) = \Delta^n f(x), \tag{1.22}$$

the nth differences. These differences may be set out as follows:

x	$y = f(x)$				
0	$f(0)$				
		$\Delta f(0)$			
1	$f(1)$		$\Delta^2 f(0)$		
		$\Delta f(1)$		$\Delta^3 f(0)$	
2	$f(2)$		$\Delta^2 f(1)$		$\Delta^4 f(0)$
		$\Delta f(2)$		$\Delta^3 f(1)$		$\Delta^5 f(0)$
3	$f(3)$		$\Delta^2 f(2)$		$\Delta^4 f(1)$

from which it may easily be seen that in order to express $f(3)$, say, in terms of the leading differences $\Delta f(0), \Delta^2 f(0), \Delta^3 f(0), \ldots$, a sequence of substitutions is necessary which precisely parallels the construction of the Figurate Triangle using the addition rule (1.6). Thus we have

$$\begin{aligned}
f(3) &= f(2) + \Delta f(2) \\
&= f(1) + 2\Delta f(1) + \Delta^2 f(1) \\
&= f(0) + 3\Delta f(0) + 3\Delta^2 f(0) + \Delta^3 f(0),
\end{aligned}$$

and, in general,

$$f(x) = f(0) + x\,\Delta f(0) + \frac{x(x-1)}{2!}\Delta^2 f(0)$$
$$+ \frac{x(x-1)(x-2)}{3!}\Delta^3 f(0) + \ldots \Delta^x f(0), \tag{1.23}$$

the series terminating earlier if zero differences are encountered. The analogy between the formation of the coefficients and the formation of the figurate numbers by the addition rule is complete, as may best be seen by arranging (1.22) in the form

$$\Delta^{n-1}f(x + 1) = \Delta^{n-1}f(x) + \Delta^{n}f(x) \qquad (1.24)$$

to parallel the addition rule. This analogy lends itself to symbolic representation through which books on finite differences develop the theory using the algebra of binomial expansions [30]. But for our purposes it is sufficient to note that as early as 1611 Harriot knew of the intimate relation between the figurate numbers and the forward-difference formula which was to attract the names of Newton and Gregory when they discovered it in the 1670's [31]. In due course that formula itself was to lead to Taylor's theorem [32], sprinkling the figurate numbers throughout analysis.

In 1615 Johann Faulhaber, of Ulm, gave equation (1.18) in a figurate context [33], and then in 1636 Pierre de Fermat, evidently not conversant with the work of either Faulhaber or Briggs, was delighted to discover equation (1.17). The result came to him whilst reading Diophantus of Alexandria's *Book on Polygonal Numbers*. This book (*ca* A.D. 250) contains at one place [34] the formula relating a polygonal number to the number of sides, and the number of points in each side, of the figure that generated it. In the margin of this section Fermat jotted down [35] a "beautiful and wonderful proposition that I have discovered", namely that "taking the natural numbers, the product of one of them with its successor gives twice the triangular number corresponding to the initial number; taking one of the natural numbers and multiplying it by the triangular number corresponding to its successor gives three times the pyramidal number corresponding to the initial number; indeed, taking one of the natural numbers and multiplying it by the pyramidal number corresponding to its successor, one obtains four times the triangulo-triangulaire of the original number. *And so on indefinitely by this general rule*". Fermat gave no proof, arguing (as was his wont) that the margin of his Diophantus was too small to contain it. Since the result was new to him we can only suppose that he had not read, or at least not remembered, Cardano's version, for he could hardly have been oblivious to the identity of the figurate and combinatorial numbers. Possibly he had been inspired by the

Arithmetic of Maurolico published in 1575 (though written in 1557 [36]) in which the author had given the classical rule (1.4) in the form "Each natural number multiplied by its successor gives twice the corresponding triangular number" [37]; the demonstration was that of Fig. 6.

Fermat's pleasure was as much to do with the use of the proposition in the summation of the powers of the integers [38] as with the beauty of the result itself. He first gave it in a letter to Father Marin Mersenne written in September or October 1636 [39] and subsequently in a letter to Gilles Personne de Roberval dated 4 November 1636 [40], where he added "All these propositions, however pretty in themselves, have aided me in the quadrature that I am pleased you value".

Summation of the powers of the integers was achieved by substituting (1.18) in the right-hand side of (1.7) to give

$$\sum_{i=1}^{l} \frac{i(i+1)(i+2)\ldots(i+k-2)}{(k-1)!} = \frac{l(l+1)(l+2)\ldots(l+k-1)}{k!}.$$

(1.25)

Now the summand on the left is a polynomial whose highest-order term is i^{k-1}, so that the whole equation can be rearranged to give the sums of the $(k-1)$th powers of the integers in terms of the sums of the $(k-2)$th and lower powers. For example, we may recover the classical result of Archimedes on the sum of the squares by putting $k = 3$ in (1.25) to find

$$\sum_{i=1}^{l} i^2 = \frac{l(l+1)(l+2)}{3} - \sum_{i=1}^{l} i = \frac{l(l+1)(2l+1)}{6}, \quad (1.26)$$

since the sum of the first l integers is $\frac{1}{2}l(l+1)$. Such results were essential for Fermat's work on the quadrature of curves, and he obtained them in ignorance of the fact that by 1631 Faulhaber [41] had published not only all the formulae for the sums up to the seventeenth powers, but had also indicated a general procedure.

By Pascal's time, therefore, the study of the figurate numbers alone had produced a Triangle capable of indefinite extension in both directions, with several general relations amongst the numbers known, and many applications realized. Their equivalence to the binomial numbers had been explicitly recognized. With the *Traité* of Pascal the figurate numbers lost their identity and became merged with the binomial coefficients. Equation (1.18) ousted (1.7) from the textbooks—multiplication was preferred to addition. As we shall see, the numbers lingered on in the simultaneous *Arithmetica infinitorum* of John Wallis, who interpolated intermediate values, and finally expired in Bernoulli's *Ars conjectandi* whilst giving birth to the Bernoulli numbers [42].

We shall cover these developments in later chapters, and close our account of the figurate numbers by noting Eastern sources which, though they had no known influence on the more fruitful European development, gave some of the results we have quoted, but at an earlier date. It is reported [43] that the formula $\frac{1}{2}l(l + 1)$ for the sum of the first l natural numbers, giving the lth triangular number (equation 1.4), is very old in Hindu writing; the sum of the first l triangular numbers to form the lth tetrahedral number is given as

$$f^l_3 = \frac{l(l + 1)(l + 2)}{6} = \frac{(l + 1)^3 - (l + 1)}{6} \tag{1.27}$$

in A.D. 499 by Aryabhata I, and the general formula (1.18) is given in the *Ganita Kaumudi* of Narayana (1356). The *Ganita Kaumudi* was a commentary on the *Lilavati* of Bhaskara [44], which contained combinatorial formulae (as we shall see in Chapter 3), and Narayana noted the equivalence of the figurate numbers and the formula for the number of combinations of different things taken so many at a time. Bhaskara himself appears not to have noticed this, even though he gives our Fig. 8 (without the first row, but extending to the ninth column) immediately after the section on combinations [45]. Although interpolation using first and second-order differences is recorded in the work of Brahmegupta (A.D. 665), there appears to have been no further development in India, and no recognition of the role of the figurate numbers in the theory [46].

In China, Yang Hui (1261), who also depicted the Binomial Triangle up to the sixth power (see our Chapter 5), is reported as possessing the first form of equation (1.27) [47], whilst a little later Chu Shih-chieh in his *Precious Mirror of the Four Elements* is variously reported as giving, without proof, equation (1.25) for k up to six [48] or for any k [49]. These series arose in a figurate context, since Chu Shih-chieh was concerned with piles of balls arranged in triangles, pyramids, and other figures, but as he was also concerned with the binomial expansion and finite differences only a re-examination of the original text in the light of the distinction between the figurate and binomial numbers will tell exactly what he achieved [50]. His determination of the value of a function whose first differences are the cubes of the integers by using Newton–Gregory forward differences [51] is undoubtedly extremely impressive, but the question of the identification of the coefficients with the figurate numbers remains.

Very similar comments could be made about Arabic writers from the time of Alhazen and Al-Karaji (*ca* 1000) to that of Al-Kashi (died 1429). As we see in Chapter 5, knowledge of the binomial expansion probably entered Europe from this source, but whether anything like equation (1.25) did is doubtful. It is, however, certain that Al-Kashi knew an explicit rule for summing the fourth powers of the integers [52].

Notes to Chapter 1

[1] Heath (1921) *Greek Mathematics*, Vol. I (*From Thales to Euclid*), pp. 76–84; Dickson (1920), pp. 1–2.
[2] Nicomachus of Gerasa (100).
[3] Gillings (1981).
[4] Gillings (1981), p. 693.
[5] See especially note 14 to Chapter 4.
[6] Stifel (1544), foglio 44*v* & 45*r*; Stifel (1545), foglio 71*v*.
[7] Smith (1925), p. 148.
[8] Smith (1925), p. 508, represents Stifel's figure incorrectly, whilst Rouse Ball (1908), pp. 216–17, describes his contribution wrongly.
[9] Scheubelius (1545), *Tractatus quintus* (pages not numbered).
[10] Vieta (1591), pp. 294–7, in *Ad Angulares Sectiones*.
[11] The best direct account is in the footnotes by D. T. Whiteside to Newton's notes: Newton (1664), pp. 81–2.
[12] Vieta (1591), p. 295.
[13] Vieta (1591), pp. 16–19, in *Ad Logisticen Speciosam*.
[14] Mahoney (1973), ch. II, pp. 26–71.
[15] Oughtred (1631), p. 48.
[16] There is a short biography of Oughtred by Cajori (1916).

[17] Oughtred (1652); now in the Library of Trinity College, Cambridge (see Whiteside, 1967, p. 22).

[18] Briggs (1633).

[19] Briggs (1633), ch. VIII, pp. 20–28; the *Abacus* is on page 21.

[20] Hutton (1785), pp. 73–81, gives a description of Briggs' *Trigonometria Britannica*.

[21] Hutton (1786), pp. 65–82, pursues his advocacy of Briggs a little too enthusiastically, for he mentions Cardano's Combinatorial Triangle but overlooks the associated result.

[22] Accoungs may be found in Goldstine (1977), pp. 13–32; Whiteside (1961a), pp. 232–6; Turnbull (1933), pp. 164–70; Hutton (1785), pp. 60–81.

[23] Briggs (1624).

[24] Whiteside (1961a), p. 233.

[25] Whiteside (1961a), p. 235.

[26] British Library Add. 6782, described in part by Lohne (1965, 1979). D. T. Whiteside has kindly furnished me with additional information.

[27] Whiteside (1971), p. 4.

[28] Goldstine (1977), p. 26, suggests "that Briggs, upon his arrival at Oxford, learned of this interpolation formula from Harriot, since the two overlapped by about two years". Briggs took up his duties as Savilian Professor in Oxford in January 1620, whilst Harriot, his predecessor, died in London on 2 July 1621, having lived there for some time. Goldstine's remark seems based on the misapprehension that Harriot lived in Oxford. Wallis (1685a), in the Preface to his *Algebra*, mentioned that Harriot "left many good things behind him in writing. Of which there is nothing hitherto made publick, but only his *Algebra* or *Analytice*".

[29] Lohne (1972), p. 125.

[30] For example Jordan (1965).

[31] Edwards (1979), pp. 283–5.

[32] Edwards (1979), p. 287; Struik (1969), pp. 328–33, translates the relevant part of Brook Taylor's *Methodus incrementorum directa et inversa* (London, 1715).

[33] Tropfke (1924), p. 22. The reference is to the anonymous *Mysterium Arithmeticum*, usually attributed to Faulhaber (Faulhaber, 1615).

[34] Diophantus of Alexandria (250), pp. 290–1.

[35] Footnote to page 291 of Diophantus (250); Fermat (1636), Vol. I, p. 341 (Latin), and Vol. III, p. 273 (French).

[36] Smith (1923), p. 302.

[37] Maurolico (1575), Book I, p. 5, Proposition 7.

[38] Mahoney (1973), pp. 229–32.

[39] Fermat (1636), Vol. II, p. 70; Itard (1948) argues in favour of 1638.

[40] Fermat (1636), Vol. II, pp. 84–5.

[41] Faulhaber (1631), pages not numbered; see our Chapter 10 for a fuller account.

[42] Or so I thought until I encountered a book (Joncourt, 1762) devoted entirely to the triangular numbers, with a table up to $n = 19999$.

[43] Singh (1936).

[44] Srinivasiengar (1967), p. 94.

[45] Bhaskara (1150), p. 52.

[46] Srinivasiengar (1967), pp. 67–8.

[47] Ho Peng-Yoke (1976), p. 540.

[48] Ho Peng-Yoke (1971), p. 268.

[49] Needham (1959), pp. 138–9, apparently quoting Sarton (1947), Part I, p. 702; the attribution is queried by Ho Peng-Yoke (1971), pp. 268–9.

[50] It is practically useless trying to elucidate such matters from secondary sources. It appears that, despite efforts mentioned by Needham and Sarton, no English translation of the *Precious Mirror* has yet been published. A special danger arises when verbal rules are summarized by modern formulae, into which more may be read than is justified. Equation (1.25) is a particularly good example, in both its Western and Eastern contexts. As it stands it looks very impressive, but all it really says is our equation (1.7), which is essentially axiomatic for the figurate numbers given their geometrical construction, plus (1.18). Now either Chu did possess (1.18) as a rule enabling any figurate number to be constructed by multiplication rather than by the defining addition process, or he did not. If he did not, representing his knowledge by the modern form (1.25) is misleading, but if he did, it is the *right-hand side* of (1.25), that is, the form (1.18) itself, which is interesting, rather than the whole equation, for the reason just given.

One recent scholar who is familiar with Chu's text, Hoe (1977), pp. 300–21, states that Chu gave, with names, the first five series of figurate numbers, starting with the natural numbers, and that "the solutions given [to various problems] show that he knew the sum of each of these five series"—but in what form? Hoe admits to being uncertain not only whether Chu possessed the formula for the general sum, but even whether he knew that each series gives the sum of its predecessor (our 1.7); the resolution of the question apparently turns on the interpretation of a particular Chinese phrase, and, that being so, we had best leave it.

As we shall see in Chapter 3, the Hindus possessed (1.7) in a combinatorial context whilst the Arabs possessed it in a binomial context (Chapter 5).

[51] Ho Peng-Yoke (1971), pp. 269–70.

[52] Smith (1925), p. 505.

2

Three combinatorial rules

In how many ways can two things be chosen from n different things? In the course of his *Isagoge*, or introduction, to Aristotle's *Categories*, Porphyry [1] (*ca* A.D. 275) needed to know how many paired comparisons could be made amongst the five Aristotelian "voices" of *genus*, *species*, *proprium*, *differentia* and *accidens*, since he wanted to comment on all the similarities and differences between them. He correctly argued that first appearances are deceptive, and that there are not simply twenty differences arising from the fact that each of the five can be paired with each of the remaining four: "Such is not the case; for though the first of the five can be paired with the remaining four, when we come to the second, one of the pairs will already have been counted; with the third, two; with the fourth, three; and with the fifth, four; thus there are in all only ten differences: $4 + 3 + 2 + 1$".

Since Porphyry arrived at the correct answer by a general argument, and not only by enumeration, we may allow him the knowledge that the number of unordered pairs that can be chosen from n things is given by the $(n - 1)$th triangular number, our f_2^{n-1}. The general solution was certainly known to Pappus [2] only a few years later (*ca* A.D. 320). He was considering a problem involving n distinct lines, no two of which are parallel and no more than two of which intersect in the same point. Each pair of lines defines a point: how many such points are there? The triangular numbers provide the answer, said Pappus, and we may suppose that he knew the reason: adding each line creates as many new points as there are existing lines, so there must be $1 + 2 + 3 + \ldots + (n - 1)$ points in all.

Boethius (*ca* A.D. 510) wrote two versions of a commentary on the

Isagoge of Porphyry, and in the second [3] he not only gave Porphyry's own reasoning but added the argument that since, respective of order, two things can be chosen from n in $n(n-1)$ ways, and each pair then occurs in two different orders, the number of combinations must be $n(n-1)/2$, thus confirming the fact that $f_2^{n-1} = n(n-1)/2$ (with which he will have been familiar, since he wrote an *Arithmetic* based on Nicomachus's work).

Pappus's solution, given above, is contained in a commentary on Euclid's lost book on *Porisms*. Pappus first noted that a number of the porisms could be summarized in a single proposition, namely that if of the six points of intersection of four lines three, all on one line, are held fixed, whilst two of the remaining three lie on fixed lines, the third will also lie on a fixed line. (We may note that this theorem leads to the Theorem of Desargues, a special case of Pascal's famous theorem that the points of intersection of the opposite sides of a hexagon inscribed in a conic lie on a line.)

Pappus then gave the generalization to an arbitrary number of lines (which is quite easy if they are added one at a time), observing that the number of points other than those which are held fixed and lie all on one line is a triangular number. In other words, if there are $(n+1)$ lines in all, n of them generate $\frac{1}{2}n(n-1)$ points of intersection. He then added "It is unlikely that Euclid was ignorant of this [generalization]", which makes it certain that it is not itself one of the lost Porisms (though Chasles [4] included it as Number XXX in his reconstruction of Euclid's book). We shall probably never have a better opinion than Pappus's as to whether Euclid knew that the number of combinations of n things taken two at a time is a triangular number, and we must let the matter rest there.

What is impressive about these instances of the derivation of the triangular number $\frac{1}{2}n(n-1)$ for the number of selections of two things from n different things in late classical times is the confidence with which the authors handled the problem. The accounts read as though they regarded the problem as straightforward, if not commonplace. All three authors were studied widely in the Middle Ages.

Pappus's *Collection*, which contains his commentary on *Porisms*, provides further evidence of his familiarity with combinatorial questions in the commentary on Apollonius's two lost books *On Contacts or Tangencies*. How many combinations of three *kinds* of things are there (in this case, circles, straight lines, and points) taken three at a time? Pappus writes simply "From three different kinds of things there are ten distinct unordered triplets", and he proceeds to list them [5]. A little later [6] he points out that six unordered pairs can similarly be formed, and although he gives no general formula, there is again an air of confidence in the way these results appear as unremarkable lemmas required for geometrical theorems. It is pleasing to believe that Pappus would not have overlooked the fact that 10 and 6 are triangular numbers (especially as the answer to "how many singles?" is so obviously 3) and that he did not give the general result simply because he did not need it. We shall probably never know.

The classical "$\frac{1}{2}n(n-1)$" rule for the number of combinations of n things taken two at a time was joined in the sixteenth century by two other rules, $2^n - 1 - n$ for the number of combinations of n things taken two or more at a time, and $n!$ for the number of permutations of n things. These rules were regarded as the basis of combinatorial theory until they were subsumed into the developing modern theory in the seventeenth century.

The precursor to the "$2^n - 1 - n$" rule was first given by Cardano in 1539 in his *Practica Arithmetice* [7] in the following form. To find the number of combinations of n things taken two or more at a time, write down the first n terms of the geometric series $1, 2, 4, 8, 16, \ldots$ and find their sum, from which n is then to be subtracted to give the answer. Cardano did not remark that the sum of the series was $2^n - 1$. He worked out the cases $n = 7, 11,$ and 22 correctly, but gave no reasoning (we of course can see that the total number of combinations is 2^n, since each thing can be taken or not, of which 1 is the combination of nothing taken and n are the single combinations, giving $2^n - 1 - n$ combinations of two or more, in accordance with the rule). Then Stifel (1544) [8], who says he first encountered this rule in Cardano's book, used it to determine first, how many new numbers could be produced by multiplying some or all of four prime numbers, 2, 3, 5 and 7, together, and, secondly, how many divisors ("aliquot parts") the product of all four, 210, has. The answer to the first question is $1 + 2 + 4 + 8 - 4 = 11$, and Stifel lists the eleven numbers, whilst the answer to the second is found by a variant of the rule in which the number of original numbers, 4, is not subtracted, giving simply $1 + 2 + 4 + 8 = 15$ as the number of divisors of 210.

These are also listed (including 1 but excluding 210 itself). As a further example he finds that $2.3.5.7.11 = 2310$ has $1 + 2 + 4 + 8 + 16 = 31$ aliquot parts. "And so on" [9].

These rules were then given (also in connection with numbers) by Buckley in 1567, whom Wallis [10] reprimanded for not mentioning that the original numbers must be unrepeated primes. Buckley gave both of them in the same context as Stifel, and his "Rule of Combinations" is indeed Stifel set in verse; I do not know whether the rest of his *Memorable Arithmetic* [11] comes from Stifel as well. David [12] offers a translation of *Memorable Arithmetic* but the rendering of the Rule of Combinations is unintelligible; fortunately she also gives [13] Wallis's translation [14], which is much better, provided it is borne in mind that Buckley himself gave no numerical example. Wallis listed the combinations of five things taken one, two, three, four, or five at a time, which Buckley did not do [15].

Cardano returns to the question in his *De proportionibus* of 1570, but here he simply sums the combinatorial numbers ^nC_r and notes that the sum is a power of two, thus shortening his rule to the form $2^n - 1$ for the number of combinations of n things taken one or more at a time; again he gives no proof, although he does imply that the rule was originally his discovery, prompted by the need for an alternative to summing the combinatorial numbers: "*Et ideo inveni hanc, quae facillima est: . . .*". [16].

Unfortunately the only modern book that covers this part of Cardano's work, Maistrov's *Probability Theory: A Historical Sketch* [17], appears (in the English translation at least) to attribute the form $2^n - 1$ to Stifel: "referring to Stifel, he [Cardano] writes down the binomial coefficients and asserts, without a proof, that the number of all possible combinations of n elements is $2^n - 1$." Here Maistrov has confused the two separate occurrences of the Arithmetical Triangle in *De proportionibus*. The first, in Proposition 137, is indeed a repeat of Stifel's table, attributed to him, but is nothing to do with combinatorial questions; it is followed, rather, in Propositions 138 and 139, by an account of the application of the coefficients to the extraction of roots, just as given by Stifel in *Arithmetica Integra*. The second occurrence of the Arithmetical Triangle, this time directly concerned with combinatorial questions, is in Proposition 170, and it is here that Cardano gives his 1539 rule in its new form, without referring to Stifel (naturally, since Stifel obtained the rule from Cardano in the first place).

Presumably the justification for the original form was the pattern of enumeration

$$
\left.
\begin{array}{llllllll}
a & & & & & & & \\
b & ab & & & & & & \\
c & ac & bc & abc & & & & \\
d & ad & bd & abd & cd & acd & bcd & abcd \\
\cdot & \cdot & \cdot & & & & &
\end{array}
\right\}
\qquad (2.1)
$$

in which each new thing (here represented by a letter) is combined with all the earlier combinations as well as being taken by itself, leading to the series $1, 2, 4, 8, \ldots$. This is the pattern of enumeration given by Van Schooten [18] in 1657; we might now think of it as labelling the corners of a hypercube.

The "$n!$" rule giving the number of permutations, or arrangements, of n different things, is very much older. The particular case $n = 6$ is reported by Chakravarti [19] as occurring in a work of the Jains in India about 300 B.C., and by the time of Bhaskara (1150) the general solution is standard [20]. Further to the West, the solution is practically explicit in the Hebrew *Book of Creation* written at the end of the eighth century [21], and the commentary by Donnolo (913–970) gives a proof of the rule: "The first letter of a two-letter word can be interchanged twice, and for each initial letter of a three-letter word the other letters can be interchanged to form two two-letter words—for each of three times. And all the arrangements there are of three-letter words correspond to each one of the four letters that can be placed first in a four-letter word: a three-letter word can be formed in six ways, and so for every initial letter of a four-letter word there are six ways—altogether four times six making twenty-four words . . . and so on." The rule is given by Levi ben Gerson (1321) [22], and when it enters the Renaissance arithmetics of Pacioli, Tartaglia and Buteo [23] it does so in the form of solving the question of how many ways ten diners can be seated at a table. It is given by Wallis [24], who had it from Vossius's *De scientiis mathematicis*. The *Book of Creation* also contains some particular solutions to the problem of selecting two things from n, and the "$\frac{1}{2}n(n-1)$" rule itself is given in essence in the commentary by Donnolo.

These three combinatorial rules, "$\frac{1}{2}n(n-1)$", "$2^n - 1 - n$" and "$n!$", were the culmination of the classical and medieval combinatorial tradition, though they were soon to be swamped by more advanced rules which may have flowed from sources in the East. Typical of works which preserve their flavour is a commentary by Clavius [25] dated 1585 on the *Sphere* of Sacrobosco (born *ca* 1200; a Yorkshireman from Halifax or *Sacrobosco*), in which Clavius, who also published an arithmetic, an algebra, and a commentary on Euclid, gave the three rules. He referred to Porphyry for the "$\frac{1}{2}n(n-1)$" rule, and in connection with the "$2^n - 1 - n$" rule noted that the sum of the first n powers of 2 was $2^n - 1$, thus simplifying the rule as Cardano had done, although as late as 1634 it was given in its original form by Hérigone [26]. Mersenne, from whom Pascal probably learnt of the Arithmetical Triangle (as we shall see), wrote a chapter on combinations in his *La verité des sciences* [27] (1625) which consisted of nothing but these three rules taken from Clavius [28]. By 1636, however, Mersenne had learnt the more advanced rules then circulating in the West which we will describe in Chapter 4. For their origin we must turn to the East, where, in India, they had been developed by the middle of the twelfth century, though whether they were transmitted westward or independently rediscovered is at present uncertain.

Notes to Chapter 2

[1] Porphyry (275), p. 41; Porphyry wrote in Greek, and the Latin forms we have used come from Boethius (see note 3 below).
[2] Pappus of Alexandria (320), pp. 488–90.
[3] Boethius (510), pp. 317–25.
[4] Chasles (1860), p. 130.
[5] Pappus (320), p. 484.
[6] Pappus (320), p. 485.
[7] Cardano (1539), ch. LI, p. 73.
[8] Stifel (1544), foglio 101 *r* & *v*.
[9] Dickson (1919), Vol. I, p. 51, says that Stifel proved Cardano's rule at this point, but this is not so; the date of his reference to Cardano is also wrong.
[10] Wallis (1685b), p. 114.
[11] Buckley (1567).
[12] David (1962), pp. 181–91.

[13] David (1962), p. 124.

[14] Wallis (1685b), p. 114.

[15] David (1962), p. 62, states that Buckley found the number of combinations of n things taken two at a time, but this is not so: it was two *or more*, and his method did not involve finding the answer for two.

[16] Cardano (1570), pp. 185–7. See our Chapter 4 for a fuller description. Thus does Cardano end up with two solutions to the original question, the equivalence of which is tantamount to the identity $1 + 2 + 4 + \ldots + 2^{n-1} \equiv 2^n - 1$, a result as old as the Babylonians (see Van der Waerden, 1981, p. 670).

[17] Maistrov (1974), p. 24.

[18] Van Schooten (1657), pp. 373–80; see also Todhunter (1865), p. 30.

[19] Chakravarti (1932), p. 86.

[20] Bhaskara (1150), Rule, p. 123.

[21] Rabinovitch (1973), pp. 143–5; Rabinovitch (1970a), pp. 203–4, from which the quotation is also taken. The dating is due to Waxman (1960), p. 388.

[22] Levi ben Gerson (1321), p. 84 of the German translation.

[23] Pacioli (1494), foglio 43v & 44r; Tartaglia (1556), foglio 16v & 17r; Buteo (1559), pp. 219–21, where there are twelve diners. Maistrov (1974), p. 25, says "Tartaglia, however, does not prove this rule in its general form", a conclusion with which I disagree; in any case Tartaglia was repeating Pacioli practically word for word. Maistrov (or perhaps his English translator, for the original work is in Russian) renders the Italian *pasti* as "dishes" rather than "meals", thus raising the rather alarming image of ten people being "served as many dishes as there are different ways in which they can be seated"!

[24] Wallis (1685b), pp. 115–17.

[25] Clavius (1585), pp. 33–9.

[26] Hérigone (1634), *Arithmétique practique*, pp. 122–3. See also note 16 above.

[27] Mersenne (1625), pp. 527–43: Book III, ch. IX.

[28] Coumet (1972).

3

The combinatorial numbers in India

Perhaps eight hundred years before Porphyry was enumerating the pairwise combinations of the five Aristotelian "voices" [1], "in the medical work of Susruta", we are told [2], "the combinations of the six tastes taken one at a time, two a time, etc., and all at a time, [were] correctly given as 63". The result was arrived at by a systematic enumeration of all the possibilities [3]. The precursor of a general rule appeared in about 300 B.C. in the *Bhagabati Sutra* of the Jains [4], and a completely general rule was given by the Jain mathematician Mahavira in *Ganita Sara Sangraha* written in A.D. 850 [5]. Since it was repeated by the famous Hindu mathematician Bhaskara in his *Lilavati* of 1150 which is more accessible to Western readers [6], we paraphrase Bhaskara's version.

Bhaskara [7] gives a rule for finding both the number of arrangements of r things of one kind and $n - r$ of another (for example, says Bhaskara, long and short syllables in prosody) or the number of selections of r things from n different things (such as tastes, in compounding a medicine). Set out the numbers from n down to 1, and divide them by the numbers from 1 up to n respectively. Then the first number obtained is the number of arrangements when there is only one of one kind of thing (and $(n - 1)$ of the other); multiply by the second number and we have the number of arrangements when there are two things of one kind (and $(n - 2)$ of the other); multiply this result by the third number and we have the number of arrangements when there are three things of one kind, and so on. Bhaskara takes $n = 6$ for two of his examples: in prosody, six syllables which are either long or short, and in medicine, six tastes which are to be combined in all possible ways (sweet, pungent, astringent, sour, salt and bitter). Applying the rule,

27

we form the sequence

$$\frac{6}{1}, \frac{5}{2}, \frac{4}{3}, \frac{3}{4}, \frac{2}{5}, \frac{1}{6}$$

and find the successive numbers to be 6, 15, 20, 15, 6 and 1 for the number of arrangements of six syllables of which 1, 2, 3, 4, 5 and 6 are long, and the remainder short, as well as for the number of medicinal preparations using 1, 2, 3, 4, 5 and 6 of the ingredients listed. To complete the solutions, Bhaskara notes that in the case of prosody, we must not forget the case of 6 short syllables, making 64 possibilities in all, whilst with the mixing of ingredients the number is 63, since at least one ingredient must be used.

No justification of the rule is given, in accordance with the usual Indian practice; there is not even an explanation of why the two problems are isomorphic. It may have been thought obvious that each selection of r things from n different things corresponds to a particular arrangement of r things of one kind and $(n - r)$ of another: for we only have to imagine the n different things of the first problem to be in order and then call the r that have been selected "things of one kind" and the $(n - r)$ left behind "things of another kind" and we have a particular arrangement in the second problem.

In modern notation, which in no way exaggerates what this "multiplicative" rule contains, the number of selections, or combinations, of r things from n different things is

$$^{n}C_{r} = \frac{n(n - 1)(n - 2)\ldots(n - r + 1)}{1 \cdot 2 \cdot 3 \ldots r}, \tag{3.1}$$

which is also the number of arrangements of r things of one kind and $(n - r)$ of another. In addition to identifying and solving these two combinatorial problems, Bhaskara gave the rows of the Combinatorial Triangle for $n = 6$ and (in another example) $n = 8$, from which it will have been obvious that

$$^{n}C_{r} = {}^{n}C_{n-r} \tag{3.2}$$

independently of any combinatorial justification. Moreover, the reason for this is clear from the operation of the rule, symmetrically placed fractions in the sequence cancelling out.

The fact that Bhaskara takes for one of his examples the old problem from Susruta of combining tastes is an interesting indication of continuity; but we are left to speculate on how the result was derived. The modern solution seems the most likely, as follows. Bhaskara devotes a whole chapter to the rules for the combination of digits [8], the first being "The product of multiplication of the arithmetical series beginning and increasing by unity and continued to the number of places will be the variations [i.e. permutations] of number with specific [different] figures" (this rule continues with a rule for finding the sum of the numbers formed by the permutations of digits, an impressive piece of number theory made possible by the Hindu method of numeration). This ancient "$n!$" rule [9] is modified when some of the digits are repeated: "The permutations found as before, being divided by the combinations separately computed for as many places as are filled by like digits, will be the variations of number", or the number of arrangements of a things of one kind, b of another, c of another, . . . , there being n all told, is

$$\frac{n!}{a!\, b!\, c! \ldots}. \tag{3.3}$$

This beautiful result is new to Bhaskara [10]. It would, of course, be an anachronism to suggest that from it Bhaskara derived his first rule by putting $a = r$ and $b = n - r$, but it does clearly exhibit the notion of division to account for arrangements rendered indistinguishable by the repetition of digits. Thus it seems not unreasonable to suppose that the rule (3.1) was actually obtained in earlier times by observing that if r things are to be selected from n different things one may start by considering the $n!$ arrangements of all the things, and then concentrating on the first r places. There will be $(n - r)!$ arrangements each starting with the same ordered selection in the first r places, because that is the number of ways the remaining $(n - r)$ places can be occupied by the remaining $(n - r)$ things. There are therefore $n!/(n - r)! = n(n - 1)(n - 2)\ldots (n - r + 1)$ different ordered selections of r things from n (a rule also given by Bhaskara but probably not his [11]), each ordered selection being one of the $r!$ arrangements of that selection; hence the general result (3.1).

Alternatively, a more direct argument might have been used. Consider Boethius's argument, given in Chapter 2, for obtaining $n(n-1)/2$ as the number of pairs that can be selected from n different things. He argued that the first thing can be chosen in n ways, and that to find the number of pairs we should first note that one from each of the remaining $(n-1)$ things can be added to each of the original n selections, making $n(n-1)$ pairs in all, but each then occurs twice. This argument is capable of natural extension. After making up $n(n-1)/2$ pairs, there remain $(n-2)$ things for selection in each case. There will thus be $n(n-1)(n-2)/2$ ways in which from n things three can be chosen as a pair and a singleton, and since the singleton can be chosen from the three things in three ways, there will be $n(n-1)(n-2)/1.2.3$ selections of three things. The argument may be continued, its essence being the relation (4.4) we meet in the next chapter; but it may be thought a little contrived.

Whatever the reasoning, we are left with the fact that this "multiplicative" rule (3.1) was available in India in A.D. 850. As we shall see, its first appearance in the West is in 1321, but it was not until 1570 that it joined up with the "additive" rule (provided by Pascal's Triangle) in the *De proportionibus* of Cardano. We now trace this rule in India and find the two rules together for the first time in 1068.

A little after the precursor of the multiplicative rule made its appearance, a writer on prosody by the name of Pingala (*ca* 200 B.C.) gave a rule, by all accounts very cryptically, for finding the number of combinations of n syllables [12] taken one at a time, two at a time, three at a time, ..., all at a time. It seems to amount to the observation that the natural numbers give the answer to the first question for successive values of n, the triangular numbers (formed by addition as described in Chapter 1) to the second question, the tetrahedral numbers to the third question, and so forth. The scheme of enumeration given in Fig. 12 (p. 35) for Rabbi ben Ezra's similar result makes this plain [13]. Naturally, any one of a number of isomorphic enumerations would serve equally well.

Pingala's rule, known as the *Meru Prastara*, is most succinctly given by his commentator Varahamihara [14] who wrote (in A.D. 505) "It is said that the numbers are obtained by adding each with the one which is past the one in front of it, except the one in the last place". A further commentator Bhattotpala (1068) has given an

	Taken two at a time	Taken three at a time	Taken four at a time
16			
15	120		
14	105	560	
13	91	455	1820
12	78	364	1365
11	66	286	1001
10	55	220	715
9	45	165	495
8	36	120	330
7	28	84	210
6	21	56	126
5	15	35	70
4	10	20	35
3	6	10	15
2	3	4	5
1	1	1	1

Fig. 11   Bhattotpala's example of the operation of the *Meru Prastara* rule for finding the combinations of sixteen different things

explicit example [15] involving combinations of sixteen things (Fig. 11). It is obvious that the rule generates the Arithmetical Triangle; it is none other than the fundamental rule of formation of the figurate numbers of Theon and Nichomachus (equation 1.5) applied without limit.

In the tenth century another commentator Halayudha [16] even gave the familiar "triangular" form of the Triangle (as in Fig. 19, page 51) but with squares instead of circles: "Draw a square. Beginning at half the square, draw two other similar squares below it; below these two, three other squares, and so on. The marking should be started by putting 1 in the first square. Put 1 in each of the two squares of the second line. In the third line put 1 in the two squares at the ends and, in the middle square, the sum of the digits in the two squares lying above it. In the fourth line put 1 in the two squares at the ends. In the middle ones put the sum of the digits in the two squares above each. Proceed in this way. Of these lines, the second gives the combinations with one syllable, the third the combinations with two syllables, etc."

No comment is needed from us on this invention of the Combinatorial Triangle six centuries, ten centuries, or even seventeen centuries (reckon it as you will) before it was used for the same purpose in the West, though we cannot overlook the reference to it by Needham [17]—"it concerns prosodic combinations only and has nothing to do with binomial coefficients", for of course the whole import of the Arithmetical Triangle is that it is to do with combinations *and* figurate numbers *and* binomial coefficients.

But to return to the joining of the two rules, the multiplicative and the additive, for finding the combinatorial coefficients. It was Bhattotpala, whom we mentioned above, who observed that the number of combinations could equally well be obtained by either rule, thus providing an alternative formula (3.1) for the coefficients obtained by the construction of the Arithmetical Triangle, whatever the context; it is interesting that the context for this identity was combinatorial in both India and the West, but binomial in Persia, as we shall see in Chapter 5 [18].

We have already noted at the close of Chapter 1 that Narayana commented on the identity of the figurate numbers and the combinatorial coefficients in 1356. All that the Hindus lacked was the appreciation that the coefficients were also those that occur in the binomial expansion. Brahmegupta [19] got as far as $(a + b)^3$ in A.D. 628, which was one power further than Euclid, but that was that. The Arithmetical Triangle still awaited its Pascal, and he was not to be found in India.

Notes to Chapter 3

[1] See Chapter 2.
[2] Chakravarti (1932), p. 79.
[3] Chakravarti (1932), pp. 81–2.
[4] Srinivasiengar (1967), pp. 26–7.
[5] Srinivasiengar (1967), p. 73.
[6] Colebrooke (1817).
[7] Bhaskara (1150), pp. 49–50 (Chapter IV: *Investigation of Mixture*, Section VI: *Permutation and Combination*).

[8] Bhaskara (1150), pp. 123–7 (Chapter XIII: *Combination [of Digits]*).

[9] See Chapter 2.

[10] According to Srinivasiengar (1967), p. 83. It is clear from Bhaskara's example (the number of permutations of the four digits 1, 1, 2, 2, which is 4!/(2.2)) that his rule has been correctly interpreted as applying to cases where more than one digit is repeated.

[11] Srinivasiengar (1967), pp. 26–7.

[12] Chakravarti (1932), pp. 82–5; Srinivasiengar (1967), pp. 27–8.

[13] Wallis (1685b), pp. 109–14 (Chapter I: *Of the variety of Elections, or Choise, in taking or leaving One or more, out of a certain Number of things proposed*) and Montmort (1713), pp. 1–9 (Proposition I of the *Traité des Combinaisons*), both give the argument by enumeration at length.

[14] Chakravarti (1932), pp. 84–5.

[15] Chakravarti (1932), p. 85; the date is from Colebrooke (1817), p. xlvi.

[16] Chakravarti (1932), pp. 82–3; Srinivasiengar (1967), pp. 27–8. We have taken some minor liberties with Chakravarti's translation.

[17] Needham (1959), footnote on p. 137.

[18] For the difficulty of interpreting the Chinese material see note 50 to Chapter 1.

[19] Brahmegupta (628), p. 279.

4

The combinatorial numbers in the West

In the last chapter we recorded the development of combinatorial rules in India, a development far in advance of contemporary knowledge in the West. By the middle of the twelfth century much of the elementary modern theory was understood, whilst in the West the classical inheritance had only produced the "$\frac{1}{2}n(n-1)$" rule, with which was coupled the "$n!$" rule derived from eighth-century Hebrew sources, and even this had been used earlier in India (Chapter 2).

However, in about 1140 in Spain, Rabbi ben Ezra [1] used the same method as in the Hindu *Meru Prastara* for finding the number of combinations of 7 things taken r at a time. The 7 things were the 6 known planets and the sun; a tabular representation of ben Ezra's method of enumeration is given in Fig. 12 for the case $r = 3$. Other values of r were similarly treated, resulting in the combinatorial numbers $^7C_2, ^7C_3, \ldots, ^7C_7$, namely $21, 35, 35, 21, 7$ and 1, with total 120. Since ben Ezra knew the classical rule that the sum of the first k integers is $\frac{1}{2}k(k+1)$ (equation 1.4), he used this short cut rather than the additions shown, for explanatory purposes, in Fig. 12 [2]. He noted that all these coefficients (bar 1) were divisible by 7, a special case of Leibniz's general result (which we shall meet in Chapter 8) that n divides $^nC_1, ^nC_2, \ldots, ^nC_{n-1}$ if it is prime. The methodical way in which Rabbi ben Ezra arrived at the combinatorial numbers for $n = 7$ leaves little doubt that he could have handled other numbers with equal facility, but he did not give a general formulation.

Then in 1321 another Jew, Levi ben Gerson, who lived in France, wrote on permutations and combinations [3], and gave three of the principal Hindu rules: $n!$ as the number of arrangements of n things; $n(n-1)(n-2)\ldots(n-r+1)$ for the number of arrangements of n

567	467	367	267	167
	456	356	256	156
	457	357	257	157
		345	245	145
		346	246	146
		347	247	147
			234	134
			235	135
			236	136
			237	137
				123
				124
				125
				126
				127
1	1	1	1	1
	2	2	2	2
		3	3	3
			4	4
				5

$$1 \ + \ 3 \ + \ 6 \ + \ 10 \ + \ 15 \ = 35$$

Fig. 12   The enumeration of the 35 ways in which 3 planets can be chosen from 7, numbered 1234567, arranged to exhibit Rabbi ben Ezra's reasoning

things taken r at a time; and, for the number of combinations of n things taken r at a time,

$$^{n}C_{r} = \frac{n(n - 1)(n - 2) \ldots (n - r + 1)}{1 \cdot 2 \cdot 3 \ldots r}. \qquad (3.1 \; bis)$$

He commented on the fact that $^{n}C_{r} = {}^{n}C_{n-r}$. His proofs constitute the first known examples of explicit mathematical induction [4]; the first two results were formally obtained in this way, and the third derived from them by division. Thus he argued that, if P_{r}^{n} be the number of arrangements of n things taken r at a time,

$$P_{r+1}^{n} = (n - r)P_{r}^{n}; \qquad (4.1)$$

for each of the P_r^n arrangements of r things omits $(n - r)$ things, and if in each case these $(n - r)$ things are taken singly and joined to the front of the arrangement, $(n - r)$ new arrangements will result, giving $(n - r)P_r^n$ as the total number of arrangements of the n things taken $(r + 1)$ at a time; but $P_1^n = n$, whence the general result.

There is no evidence that these rules were transmitted to Europe via the Arabian route along which so much of Hindu mathematics travelled and, unless the European development was separate, which seems unlikely, other routes must be sought. An interesting thread involves Hebrew writers, as follows.

"Early in the Christian Era", writes Smith [5], "there developed a close relation between mathematics and the mystic science of the Hebrews known as the cabala. This led to the belief in the mysticism of arrangements and hence to a study of permutations and combinations. The movement seems to have begun in the anonymous *Sefer Jezira* (*Book of Creation*), and shows itself now and then in later works". We encountered the *Book of Creation* in Chapter 2; in the light of it and of the contributions of ben Ezra and ben Gerson it is natural to suggest that Hebrew writers had acquired some knowledge of the Hindu combinatorial rules. The lack of direct evidence of transmission need not be surprising if we compare it with the similar lack of evidence for the circulation of the Hindu rules in Europe in the two centuries between their description by ben Gerson and their reappearance in the sixteenth century.

However, in the West the principal spur to combinatorial enumeration in the Middle Ages was gaming with dice, and Kendall [6] has described several examples of the correct enumeration of the ways in which two or three dice can fall. He also gives the text of part of the poem *De Vetula*, apparently thirteenth-century, in which the throws of three dice are enumerated:

"If all three numbers are alike there are six possibilities; if two are alike and the other different there are 30 cases, because the pair can be chosen in six ways and the other in five; and if all three are different there are 20 ways, because 30 times 4 is 120 but each possibility arises in 6 ways. There are 56 possibilities".

This is a correct enumeration of the throws, order being ignored, but it is also possible to take order into account:

"If all three are alike there is only one way for each number; if two are alike and one different there are three ways; and if all are different there are six ways. The accompanying figure shows the various ways".

The "accompanying figure" is given in extended form in Fig. 13.

Unfortunately this method of enumeration, in which the throws are classified in the first instance by the number of points scored, does not reveal the close connection with the figurate numbers. 56 is a figurate number, and indeed the first six numbers in the final column are triangular numbers, but the series is then broken because in enumerating the throws which give nine points "117" is, of course, not allowed because there is no "7" on a die. (The general solution was to be given by De Moivre in 1711, though it was known to Montmort in 1710 [7].) Nor is there any recognition that the total number of ordered throws, 216, is simply 6^3.

The crucial analysis of dice-throws which at one and the same time solved with complete generality the number of unordered throws of n dice (even 10 000, said the author) and extended the table of figurate numbers to more than three dimensions, was made by Tartaglia "on the first day of Lent, 1523, in Verona", as he proudly tells us, "having thought about the problem all night". It is described in the second part of his *General trattato di numeri, et misure* [8] published in 1556 (Fig. 14). Tartaglia mentions no antecedents and gives no reasoning behind his solution (about which, he says, it would be necessary to write a book); his priority in this matter has not been challenged, not even by his contemporary and antagonist Cardano [9]. We may note that this enumeration problem is an extension of the one solved correctly by Pappus and considered in Chapter 2: "From three different kinds of things there are ten distinct unordered triplets".

The problem suggested itself to Tartaglia as he was watching some people casting their own horoscopes by means of a *Book of Fortune*, in which apt verses were selected by a process which included noting the numbers on the faces of three dice. The particular book Tartaglia saw was Lorenzo Spirto's *Libro della Ventura*, which had existed in various editions since 1484. The 56 ways in which three dice can fall were set out on each page; an edition of 1535 [10] has twenty such pages (one for each type of question posed) actually depicting the dice throws. The

Number of points	Corresponding throws*	Number of unordered throws	Number of permutations of each	Number of ordered throws
3 or 18	111	1	1	1
4 or 17	112	1	3	3
5 or 16	113	2	3	6
	122		3	
6 or 15	114		3	
	123	3	6	10
	222		1	
7 or 14	115		3	
	124	4	6	15
	133		3	
	223		3	
8 or 13	116		3	
	125		6	
	134	5	6	21
	224		3	
	233		3	
9 or 12	126		6	
	135		6	
	144	6	3	25
	225		3	
	234		6	
	333		1	
10 or 11	136		6	
	145		6	
	226	6	3	27
	235		6	
	244		3	
	334		3	
		28		108
× 2		56		216

*Only listed for the smaller of the number of points.

Fig. 13 An extended version of the figure from *De Vetula*. The original (which contains two errors) consists of columns one, three, and five only, and gives no totals

ne multiplica 362880. per 10 faranno 3628800. & in tanti modi potranno federe 10 perfone , & tanti paftigli fara quello albergatore a foldi vno per pafto. Farai poi conto quante lire montaran-no.&c.Et l'ordine di quefto procello renirai fe foffero ben mille perfone,o quante fi vogliano , per-che in infinitum regula tendit . Onde poi a fapere di 11 perfone multiplicarefti li modi delle 10 per-fone per 11.dicendo 11 fia 3628800.fanno 39916800.& in tanti modi potranno federe 11 perfo-ne , & per 12. dirai 12 fia 39916800. faranno 479001600. & cofi potrai procedere in infinito , il qual modo fi proua per la deduttione,che prima cominciaffimo a fare.&c.

Regola generale dal prefente auttore ritrouata il primo giorno di quarefima

l'anno 1523. in Verona , di faper trouare in quanti modi puo variar il getto
di che quantita di dati fi voglia nel tirar quelli.

Tantiando l'anno 1523 in Verona,& il giorno di Carneuale vna cometiua di giouinet-ti,et altri di matura età tracuano con 3 dati ful libro(detto)della ventura di Lorenzo fpir-to,cercando ciafcun di loro da intendere quello che tal libro gli determinaua circa alle ma-terie,che tal libro prepone da notificarli . Et vedendo che in ogni carta li detti 3 dati con la ifperienza hauea il detto auttore trouato poter variar in 56 modi,laqual cofa confiderando delibe-rai di voler trouare , come che con regola generale tal cofa fi poteffe determinare , & non folamente in detti 3 dati,ma in ogni altra maggior quantita di dati,& cofi tutta la notte fopra tal materia andai tanto freneticando che il giorno feguente(che fu il primo di quarefima) trouai tali ordini,ouer re-gole formarfi da ftrane forte di progreffioni,come intenderai.

Prima eglie manifefto che vn fol dato puo variar in 6 modi per effer di 6 fazze,ouer di 6 bafe,nellequa li fono 6 ordini di numeri,cioe 1.2.3.4.5.& 6.come in figura vedi.

Ma per trouar in quanti modi puo variar il getto di duoi dati trouai che raccogliendo tutte le vnita,che fono da 1 per fino in 6.nella fopra notata progreffion continua , che fanno 21. & cofi in 21 modo trouai poter variar il getto di duoi dati.

Tre dati poi ponno variar il lor getto nella fumma di quefti 6 termini di progreffione 1.3.6.10.15.21. laqual fumma fara 56.

Li 4 dati ponno variar il lor getto nella fumma di quefti altri 6 termini di progreffione 1.4.10.20.35. 56.laqual fumma fara 126.

Li 5 dati ponno variar il lor getto nella fumma di quefti altri 6 termini di progreffione 1.5.15.35.70. 126.laqual fumma fara 252.

Li 6 dati ponno variar il lor getto nella fumma di quefti altri 6 termini di progreffioni 1.6.21.56.126. 252.laqual fumma fara 462.

Li 7 dati ponno variar il lor getto nella fumma di quefti 6 termini di progreffione 1.7.28.84.210.462. laqual fumma fara 792.

Li 8 dati ponno variar il lor getto nella fumma di quefti 6 termini di progreffione 1.8.36.120.330. 792.laqual fumma fara 1287.

Ma a volerti mo dichiarare minutamente in fcrittura l'origine di tutti li fopra notati 6 termini di pro-greffioni bifognaria formarui fopra vn libro,ma accioche in parte refti fatisfatto,fappi che ogni vna di dette progreffioni fi forma dalla progreffione

anciana,& la prima progreffione viene a effer di 6	per 1 dato	1	1	1	1	1	1
termini di vna vnita per termine in quefta forma	per 2 dati	1	2	3	4	5	6
1.1.1.1.1.1.et cofi la fumma di quefti 6 termini di	per 3 dati	1	3	6	10	15	21
progreffione puo variar il getto di vn dato folo,	per 4 dati	1	4	10	20	35	56
come vedi in figura . Et nota che l'ultimo termine	per 5 dati	1	5	15	35	70	126
di ciafcuna di dette progreffioni vien a effer la fum	per 6 dati	1	6	21	56	126	252
ma della anciana progreffione , come nella figura	per 7 dati	1	7	28	84	210	462
puoi vedere,& cō tal ordine porrai faper li 10000	per 8 dati	1	8	36	120	330	792
dati in quanti modi ponno variar il lor getto.							

Il fine del primo libro.

Fig. 14 The page from Tartaglia's *General trattato* of 1556 giving the number of ways *n* dice can fall

111	116	336	553	651	541
222	221	441	554	652	542
333	223	442	556	653	543
444	224	443	661	654	531
555	225	·	·	641	532
666	226	·	·	642	521
112	331	445	662	643	431
113	332	446	663	631	432
114	334	551	664	632	421
115	335	552	665	621	321

Fig. 15   The enumeration of the 56 throws of 3 dice given in the *Libro della Ventura*

enumeration is shown in Fig. 15 (it is the same on each page except for obvious errors). We note that the enumeration is systematic, and indeed follows the outline given in the text (though not the Figure) of *De Vetula*: reading by columns, first the six cases with all the numbers alike are listed, then the thirty cases with two alike and the other different, and finally the twenty cases with the numbers all different. But this pattern does not reveal the connection with the figurate numbers any more than does *De Vetula*'s, and its use in the general case is complicated. It was a stroke of genius on Tartaglia's part that enabled him to look at the problem from a much more profitable point of view, as follows.

"A single die can fall in 6 ways, since it has six faces inscribed 1, 2, 3, 4, 5 and 6. To find in how many ways two dice can fall, we recall this progression and sum it, obtaining 21 ways. Three dice can fall in the number of ways given by the sum to six terms of the progression 1, 3, 6, 10, 15, 21, which is 56; four dice by summing 1, 4, 10, 20, 35, 56 to give 126". And so Tartaglia proceeds until he has dealt with eight dice, where he breaks off to give his general rule, illustrated by a table (Fig. 14). "Each one of the progressions is formed from the earlier progression, whilst the first progression is formed from 1, 1, 1, 1, 1, 1, so that indeed the sum of these six terms gives the number of throws with a single die. The last term in each of these progressions is the sum of the earlier progression, as may be seen in the figure".

The fact that Tartaglia offers no explanation of his discovery need not detain us long: it is prefectly straightforward once one has chosen to enumerate the possible throws in a particular way, shown

111	211	311	411	511	611
	221	321	421	521	621
	222	322	422	522	622
		331	431	531	631
		332	432	532	632
		333	433	533	633
			441	541	641
			442	542	642
			443	543	643
			444	544	644
				551	651
				552	652
				553	653
				554	654
				555	655
					661
					662
					663
					664
					665
					666

1	3	6	10	15	21

56 ways in all

Fig. 16   The enumeration of the 56 throws of 3 dice arranged to exhibit Tartaglia's presumed reasoning

in Fig. 16 for three dice, in which the classification is not by the number of points scored, as it was in Fig. 13, but by a rule of enumeration in which the digits 1 to 6 are introduced successively. No description of this array will convey its essence to the reader more effectively than a close examination of it. It seems extremely likely that Tartaglia used such arrays for his solution, since his contemporary Buteo in France used the same ordering when solving this problem: in his *Logistica* [11], published three years after the *General trattato*, Buteo gave the enumerations of the throws of 1, 2, 3 and 4 dice, together with Tartaglia's solution, although he did not add the table given in our Fig. 14. Moreover, this same ordering of the 56 throws of three dice had been given in books of divination similar to Lorenzo Spirto's since the 10th-century *Sortes apostolorum* [12]. Tartaglia's presumed reasoning runs as follows.

However many dice are thrown, place them in non-ascending order (such as 544 rather than 445). Then the first can display 1, 2, 3, 4, 5 or 6 (there are thus 6 possibilities with only one die), but for the second, not being allowed to exceed the first, there are only 1, 2, 3, 4, 5 and 6 possibilities respectively, or 21 in all. Now consider a third die. When the first displays 1, so must the second, and so must the third, producing just one possibility. When the first displays 2 there are, as we have seen, two possibilities for the second, 1 and 2, and since the third die cannot display more than the second, the total number of possibilities must be the same as the total number of possibilities for the second when the first is 1 or 2, which we have seen to be $1 + 2 = 3$. When the first displays 3 the second can display 1, 2 or 3, and the third can therefore display the same numbers as can the second when the first is 1, 2 or 3, which is $1 + 2 + 3 = 6$.

In Tartaglia's time this explanation would no doubt have continued in like manner up to the eighth die, covering many pages, but we may allow ourselves the general observation that if we know that k dice can produce K_1 throws (in non-ascending order, as above) starting with 1, K_2 starting with 2, K_3 with 3, K_4 with 4, K_5 with 5, and K_6 with 6, then adding a $(k + 1)$th die will lead to the enumeration

$$
\left.
\begin{aligned}
&K_1 && \text{throws starting with } 1 \\
&K_2 + K_1 && \text{''} \quad 2 \\
&K_3 + K_2 + K_1 && \text{''} \quad 3 \\
&K_4 + K_3 + K_2 + K_1 && \text{''} \quad 4 \\
&K_5 + K_4 + K_3 + K_2 + K_1 && \text{''} \quad 5 \\
&K_6 + K_5 + K_4 + K_3 + K_2 + K_1 && \text{''} \quad 6.
\end{aligned}
\right\} \quad (4.2)
$$

There is, of course, nothing specific to only six-sided dice in the development of the argument.

We need not be surprised that Tartaglia does not comment on the fact that the first four rows of Fig. 14 give the figurate numbers (cf. Fig. 8); he must have known it because his rule of formation is exactly the same as that of Nicomachus [13], and he had described the triangular numbers at the beginning of Book 1 of the second part of the *General trattato*. His habit of not giving cross-references even extends to omitting any comment on the fact that the numbers contained in the Binomial Triangle he gives in Book 2 (see Fig. 20, page 53) are exactly those given in Book 1 in connexion with enumerating the throws of dice [14]. It is impossible to believe he was not aware of the identity.

Thus did Nicolo Tartaglia, famous for his part in the solution of the cubic equation, contribute one of the corner-stones of combinatorial analysis, and in that connexion extend the Figurate Triangle by continued application of the classical rules for generating the figurate numbers.

To his contemporary Gerolamo Cardano we owe not only the origins of the probability calculus, but, from 1570 [15], a printed account of the number of combinations of n different things taken r at a time. He simply describes and displays a Combinatorial Triangle (Fig. 17), not many pages after having given Stifel's Figurate Triangle, and indicates how the numbers are to be used. In essence he is inviting us to subscribe to Narayana's rule (see Chapter 1):

$$^nC_r = f_r^{n-r+1}, \qquad (4.3)$$

though without any proof. But in addition he tells us how to get one combinatorial number ^nC_r from the previous one $^nC_{r-1}$:

$$^nC_r = \frac{n-r+1}{r} \, ^nC_{r-1}. \qquad (4.4)$$

This enables him to generate the successive numbers of combinations of n things taken $1, 2, 3, 4, \ldots$ at a time by repeatedly applying (4.4); or in other words he gives us the Hindu rule

$$^nC_r = \frac{n(n-1)(n-2)\ldots(n-r+1)}{1.2.3\ldots r}, \qquad (3.1 \; bis)$$

DE PROPORTIONIBVS LIB. V.   185

Propofitio centefimafeptuagefima.

Coniugationes cuiufuis numeri breuiter inuenire.

Sint gratia exempli decē homines,& patet quod poſſent eſſe ſin Co. ᵐ guli,& hoc decē modis,quia ſunt decē,ut Petrus & Ioannes : item, poſſunt eſſe omnes ſimul,& hoc uno modo tantum,& poſſunt eſſe duo,& hoc poteſt uariari q̄draginta quinꝗ modis : & poſſunt eſſe octo,& manifeſtum eſt,quod totidē modis uariantur , ſcilicet qua= draginta quinꝗ,nam cum erunt octo,duo qui relinquūtur, uariari poſſunt 45 modis,ergo & illi octo ad unguē totidem modis. Et ſi= militer tres quot modis uariantur tot modis ſeptē, & quot modis quatuor tot ſex:quinꝗ autem quia ſunt dimidium decem,pluribus modis uariantur.Et ideò pro ordine huius detrahes unū, ut ſi ſint undecim uiri pones decem,ſi decem pones noue, & colliges natu= ralem ſeriem numerorum,ut infrà uides uno ſemper termino defi= ciente:& ex priore ordine,ubi uidebis ſemper etiā duplicari nume= ros:ut 3.6.inde ſub 6.10.& 20 à latere,& ſub 20 35.& à latere 70 du=

plum 35, & ſub 70 126,& à late= re 252, & hoc p cognitione q̃d recte ſis opera= tus. Secundò a= nimaduertes ſe= quētes ordines fieri ex recta li= nea priorum, ue

1	2	3	4	5	6	7	8	9	10	11
1	1	1	1	1	1	1	1	1	1	1
2	3	4	5	6	7	8	9	10	11	
3	6	10	15	21	28	36	45	55		
4	10	20	35	56	84	120	165			
5	15	35	70	126	210	330				
6	21	56	126	252	462					
7	28	84	210	462						
8	36	120	330							
9	45	165								
10	55									
11										

lut ſextus ordo eſt 7.28.84.210.462.ita incipiendo in primo ordi= ne à 7,& tendendo ad dextram,inuenies illos eoſdem numeros ad unguem,& ita in ſeptimo ordine 8.36.120.330. à ſiniſtra inuento 8 in primo ordine , & procedendo ad dextram, inuenies 36. 120.& 330.Tertium eſt quod numeri ultimi à medio ſunt ij̄dem, ut 462 & 462.330 & 330.165 & 165.55 & 55.11 & 11.Et ſeorſum, ut dixi , rema= net 1. Oportet igitur colligere numeros angulares, ut à latere ui= des,& fit 2047 numerus coniugationum , tot enim modis poſſunt uariari.Et ſi eſſent decem tantum,ut ab initio propoſui, primus or= do finitur ad 10,ſecundus ad 45,tertius ad 120,quartus ad 210,quin tus ad 252,ſextus redit ad 210 , ſeptimus ad 120 , octauus ad 45 , no= nus ad 10,decimus ad 1. Et ita colligeretur ſumma ex extremis nu= meris angularibus 1023.Et tot erunt coniugationes.Hic uides quia numerus 10 eſt par,et quod adempta monade,relinquitur 9,qui eſt impar quòd medius qui pertinet ad quintum ordinem eſt maxi=

Q   mus,

Fig. 17   The Combinatorial Triàngle from Cardano's *Opus novum* of 1570

which, as we have seen, had been circulating in Western Europe for some time. Indeed in 1550 Cardano [16] had himself used the rule to enumerate all the ways in which $n = 20$ things may be combined (shortening his calculation by using $^nC_r = {}^nC_{n-r}$).

The interest of Cardano's account is not only that henceforth this rule was available in a standard work, but that by giving (4.4) at the same time as pointing out that the combinatorial numbers are simply the figurate numbers (4.3), he has given (1.17), that is, the same relation described for the figurate numbers, which we noticed in the *Trigonometria Britannica* of Briggs [17], and hence (1.18) also. Quite possibly Briggs obtained it from this account of Cardano's, even though he was not interested in using the figurate numbers for combinatorial purposes. But whereas Briggs explicitly connected the figurate and binomial numbers, Cardano did not, so we can only *infer* that the latter knew the binomial theorem for positive integral index [18]. However, Cardano did appreciate that the sum of the combinatorial numbers for a given n (excluding the first) was

$$\sum_{r=1}^{n} {}^nC_r = 2^n - 1, \tag{4.5}$$

a result whose history we gave in Chapter 2.

Father Marin Mersenne [19], whom we noted as having given the three classical combinatorial rules in 1625, had by 1636 learnt of everything that Cardano had written on combinations. In that year his *Harmonicorum libri XII* were published in Paris, and contained, in Book VII [20], all the principal rules, with application to finding the numbers of permutations and combinations of musical notes. In particular, Mersenne [21] gives Cardano's form of the Combinatorial Triangle (Fig. 18) but extended to a rectangle of 25 rows and 12 columns (there being no initial column of l's). The largest number is $^{36}C_{12}$, correctly given as 1251677700. The computation is by the addition rule, but then Mersenne introduces the multiplication rule (3.1) and thus derives a Table [22] giving the combinatorial coefficients for $n = 36$, $r = 1, 2, 3, \ldots, 20$, with the observation that they increase up to, and are symmetrical about, $r = 18$. They have been computed, for each r, from the second and third columns of the Table, which contain respectively the number of arrangements of r different things, $r!$, and the number of

Tabella pulcherrima & vtilißima Combinationis duodecim Cantilenarum.

	I.	II.	III.	IV.	V.	VI.	VII.	VIII.	IX.	X.	XI.	XII.
1	1	1	1	1	1	1	1	1	1	1	1	1
2	3	4	5	6	7	8	9	10	11	12	13	
3	6	10	15	21	28	36	45	55	66	78	91	
4	10	20	35	56	84	120	165	220	286	364	455	
5	15	35	70	126	210	330	495	715	1001	1365	1820	
6	21	56	126	252	462	792	1287	2002	3003	4368	6188	
7	28	84	210	462	924	1716	3003	5005	8008	12376	18564	
8	36	120	330	792	1716	3432	6435	11440	19448	31824	50388	
9	45	165	495	1287	3003	6435	12870	24310	43758	75582	125970	
10	55	220	715	2002	5005	11440	24310	48620	92378	167960	293930	
11	66	286	1001	3003	8008	19448	43758	92378	184756	352716	646646	
12	78	364	1365	4368	12376	31824	75582	167960	352716	705432	1352078	
13	91	455	1820	6188	18564	50388	125970	293930	646646	1352078	2704156	
14	105	560	2380	8568	27132	77520	203490	497420	1144066	2496144	5200300	
15	120	630	3060	11628	38760	116280	319770	817190	1961256	4457400	9657700	
16	136	816	3876	15504	54264	170544	490314	1307504	3268760	7726160	17383860	
17	153	969	4845	20349	74613	245157	735471	2042975	5311735	13037895	30421755	
18	171	1140	5985	26334	100947	346104	1081575	3124550	8436285	21474180	51895935	
19	190	1330	7315	33649	134596	480700	1562275	4686825	13123110	34597290	86493225	
20	210	1540	8855	42504	177100	657800	2220075	6906900	20030010	54627300	141120525	
21	231	1771	10626	53130	230230	888030	3108105	10015005	30045015	84672315	225792840	
22	253	2024	12650	65780	296010	1184040	4292145	14307150	44352165	129024480	354817320	
23	276	2300	14950	80730	376740	1560780	5852925	20160075	64512240	193536720	548354040	
24	300	2600	17550	98280	475020	2035800	7888725	28048800	92561040	286097760	834451800	
25	325	2925	20475	118755	593775	2629575	10518300	38567100	131128140	417225900	1251677700	

Fig. 18 The Combinatorial Triangle from Mersenne's *Harmonicorum libri XII* of 1636

arrangements of n different things taken r at a time, $n(n-1)(n-2)\ldots(n-r+1)$. Thus Mersenne, in addition to learning from Cardano, possessed Levi ben Gerson's result relating these three numbers.

In an earlier part of Book VII [23] Mersenne has inserted a section entitled *Artem Combinandi praeceptis & exemplis, aperire* written not by him but by a man he refers to as "I.M.D.M.I." (possibly Ioannes Matan [24]) and giving not only the straightforward "$n!$" rule but also extending it to handle repeats, thereby obtaining Bhaskara's Hindu rule that if one of the things is repeated a times, another b times, another c times, and so forth, the number of arrangements is

$$\frac{n!}{a!\,b!\,c!\,\ldots}. \qquad\qquad (3.3\ bis)$$

Mersenne uses this result in several places in Book VII in connection with the arrangement of musical notes ("I.M.D.M.I." had used anagrams in his examples). In England, (3.3) is given by Strode (1678) [25] and Wallis (1685) [26].

All this combinatorial material recurs in Mersenne's French version of *Harmonicorum libri XII*, the *Harmonie universelle* [27] published in two volumes in Paris in 1636 and 1637. It is interesting to note that the *Privilege du roi* for both the Latin and French works is dated 13 October 1629. In the French version Mersenne has added Buteo's result [28] from 1559 that the number of arrangements that can be made by drawing r things from n different kinds of things is n^r.

With these works the theory of combinations assumes its modern form. Not only did Mersenne bring together all the results then known, but through these volumes (famous amongst musicologists though not yet amongst mathematicians) [29], and no doubt through his numerous correspondents [30] as well, he broadcast them throughout Europe. "When Mersenne knew something, the whole of the 'Republic of Letters' was shortly informed about it" [31]. We know that the Pascals, father and son, visited him [32], so that it comes as no surprise to find that Pascal's format for the Arithmetical Triangle is the same as Mersenne's (see Chapter 6); we know that he corresponded with the Lullist Athanasius Kircherus [33], whose *Musurgia universalis* of 1650 repeats many of Mersenne's results, even to the extent of using the same numerical examples. Leibniz owned that Lullism had had some influence on his *Dissertatio de Arte Combinatoria* [34] of 1666.

The interaction of music and combinatorial theory is deepest in the English art of change-ringing bells, but this only developed from the middle of the seventeenth century (Fabian Stedman, of Cambridge, from that time, is described as the "father of change-ringing"), too late to influence combinatorial theory.

The modern neglect of the Renaissance contributions of Tartaglia, Cardano and Mersenne to combinatorial theory may be attributed to the reliance that has been placed on Isaac Todhunter's influential *History of the Mathematical Theory of Probability* published in 1865. At one point [35] Todhunter regrets the imprecision of a reference to Tartaglia given by another author, and later on [36] he fails to pursue Leibniz's references to Cardano, Buteo, Tartaglia and Clavius [37]. Mersenne is not mentioned at all. In defence of Todhunter's neglect of the early writers it may be said that the rest of his title is "... *from the time of Pascal to that of Laplace*", which not all his followers have noticed; Buckley is mentioned only because of the reference in Wallis's work [38], and it is his presence in Todhunter even though a century earlier than Pascal which has given him undue prominence (see our Chapter 2).

Notes to Chapter 4

[1] Rabinovitch (1973), pp. 145–7; Smith (1925), p. 525.
[2] Clagett (1968), p. 446, is of the opinion that ben Ezra was making use of the "$\frac{1}{2}n(n - 1)$" rule given by Boethius rather than the "$\frac{1}{2}k(k + 1)$" rule, but this seems to be a simple error.
[3] Levi ben Gerson (1321), pp. 47–55 and 84–5 of the German translation; Rabinovitch (1973), pp. 147–8; Smith (1925), p. 526. Unfortunately on p. 527 Smith attributes the rule (3.1) to Hérigone (1634), in spite of his earlier testimony.
[4] Rabinovitch (1970b).
[5] Smith (1925), p. 525.
[6] Kendall (1956); see also David (1962), pp. 31–4.
[7] Todhunter (1865), pp. 84–5.
[8] Tartaglia (1556), Part II, foglio 17r.
[9] Ore (1953) and David (1962) describe the mutual relations of Tartaglia and Cardano. Cardano (1560) himself enumerated dice throws in the manner of *De Vetula*; so also Galileo (1620).
[10] Spirto (1535). I am indebted to Mr A. J. Hunt for helping me find this.
[11] Buteo (1559), pp. 304–9, *Quaestio* 90. A later example of similar reasoning is Montmort (1713), p. 37.
[12] Braekman (1980).
[13] See our Chapter 1.
[14] Kendall (1956) is wrong to say "Not until 1556 did Tartaglia publish the scheme now known (very unjustly) as Pascal's arithmetical triangle, *and then not in a probabilistic context*" (my italics); Smith (1929), p. 86 *footnote*, says that Tartaglia pointed out the correspondence between figurate and binomial numbers, but I do not know where. Stifel, of course, knew it in 1544 (see our Chapter 1).
[15] Cardano (1570), pp. 185–7.
[16] Cardano (1550), p. 588; the rule later became known as "the rule of Hérigone" from its appearance in Hérigone (1634), *Arithmétique practique*, pp. 119–21.
[17] Briggs (1633), p. 20; an account is in Hutton (1785), pp. 73–81.
[18] I agree with Boyer (1950), p. 389: "There can be little doubt but that Cardano, like Tartaglia, was aware that the elements in the triangle are coefficients in the expansions of binomials".
[19] David (1962), pp. 196–228, gives a translation (by M. Merrington) of de Coste's *Life of Mersenne*.

[20] Mersenne (1636), pp. 113–60.

[21] Mersenne (1636), p. 136, *Propositio XII*.

[22] Mersenne (1636), p. 137.

[23] Mersenne (1636), pp. 118–19, *Propositio V*.

[24] Coumet (1972).

[25] Strode (1678), p. 14. This book contains the first English mention of Pascal's *Traité* of which I am aware, and (pp. 38–40, 44–9) a table of the figurate numbers up to f^{12}_{29} and then to f^8_{99}.

[26] Wallis (1685b), pp. 117–18.

[27] Mersenne (1637), pp. 89–180, *Livre second, des chants*.

[28] Buteo (1559), pp. 312–29. Buteo gave the result in connection with a combination lock (*Quaestio* 92) with r rings each carrying n letters; his diagram is reproduced by Smith (1925), p. 527.

[29] But see Knobloch (1979), who draws attention to Mersenne and Kircherus.

[30] David (1962), pp. 212–26.

[31] Boyer (1968), p. 367.

[32] David (1962), p. 220.

[33] David (1962), p. 224. Ramon Lull (1235–1315) believed that in every branch of knowledge there are certain principles or categories that can be assumed without question, and that knowledge is to be increased through an exploration of all possible pairwise combinations of these categories. Lull himself gave some marvellous diagrams enumerating all the combinations, but his theoretical ability seems to have been limited to a correct enumeration of pairwise combinations, for example 120 from 16 virtues (see *Histoire littéraire de la France*, Vol. XXIX, Paris, 1885, p. 75. Pages 74–9 give some account of Lull's *Ars compendiosa inveniendi veritatem, seu Ars magna et major*). Gardner (1968), chapter I, gives a brief account of Lull's life and work; Hacking (1975), p. 49, says that Lull "is usually cited as the founder of the theory of combinations", but in reality he contributed little and late. Kircherus wrote *Ars Magna Sciendi sive Combinatoria* (1669) devoted to the Lullian art, Book IV of which (Kircherus, 1669, pp. 153–201) is devoted to combinatorial theory; in *Musurgia combinatoria*, part of Book VIII of *Musurgia universalis*, he had earlier applied combinatorial theory to music (Kircherus, 1650, pp. 3–27).

[34] Gardner (1968), pp. 2–3; see our Chapter 8.

[35] Todhunter (1865), p. 1.

[36] Todhunter (1865), p. 33.

[37] The quotation from Leibniz shows, incidentally, that he looked in the wrong book of Cardano's, saying that his *Practice Arithmetice* contained nothing. In this he is not quite correct, since it actually contains the first account of the rule (4.5) which many authors, following Todhunter (1865), pp. 26–7, attribute to Buckley. Most of Cardano's combinatorial work, however, is in *De Proportionibus*. See also Hofmann (1974), p. 3.

[38] Wallis (1685b), pp. 113–14; see note 37 above.

5

The binomial numbers

When considering the figurate and combinatorial origins of the Arithmetical Triangle we separated out the Indian development because there is no external evidence that it made any contribution to the European advances which led to Pascal's *Traité* and beyond. The case of the binomial expansion is rather different. There is a clear thread running from the earliest Arab algebraists through to seventeenth-century Europe, culminating in Newton's discovery of the general binomial theorem (which we describe in Chapter 8). Essentially two things awaited discovery beyond the straightforward listing of the coefficients of the expansion of $(a + b)^n$. First, that the coefficients could be formed according to the addition rule that is one of the ways of generating the Arithmetical Triangle: using $\binom{n}{r}$ to indicate the coefficient of $a^r b^{n-r}$ in the expansion, this rule is

$$\binom{n}{r} = \binom{n-1}{r-1} + \binom{n-1}{r}. \tag{5.1}$$

Secondly, that a direct expression for this coefficient is

$$\binom{n}{r} = \frac{n(n-1)(n-2)\ldots(n-r+1)}{1 \cdot 2 \cdot 3 \ldots r}, \tag{5.2}$$

the discovery of which gives us the binomial theorem for positive integral index.

The continuity of the thread does not, of course, preclude the possibility of repeated discovery. Sarton comments [1] that "This is an excellent example of recondite knowledge that was many times discovered, lost, and rediscovered, until the matter was fully understood". In particular, the Chinese discovery of the Binomial Triangle was lost so far as the European development was concerned, and we describe it first on that account.

According to Yang Hui [2] (1261), who listed the coefficients of $(a + b)^n$ up to the sixth power, the expansion derives from the *Shih-so suan-shu* of Chia Hsien (*ca* 1100), where it is called "the tabulation system for unlocking binomial coefficients". The same triangle, extended to the eighth power, is given by Chu Shih-chieh [3] in his *Precious Mirror of the Four Elements* in 1303 (Fig. 19). The "tabulation system" is none other than (5.1), so that we may fairly attribute the Binomial Triangle to China in about 1100. As we shall see below, it is, however, at least as old in Persia. Whether Chu Shih-

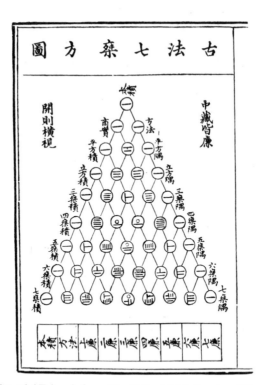

Fig. 19   The Binomial Triangle from Chu Shih-chieh's *Precious Mirror of the Four Elements* (1303), reprinted with permission from Needham (1959). Note that $\binom{7}{3}$ is given incorrectly, and should be the same character as $\binom{7}{4}$

chieh knew the general multiplicative rule (5.2) is doubtful, for the reasons we have given at the end of Chapter 1.

The Chinese development arose in connection with the solution of polynomial equations of any degree by a method closely related to Horner's [4], and although it did not influence the West, in respect of the binomial expansion it was far in advance of its Greek and Hindu counterparts. Euclid gave $(a + b)^2 = a^2 + 2ab + b^2$ in Book II of the *Elements* [5] in a geometrical context (*ca* 300 B.C.); I am not aware that any classical writer partitioned a cube in the same way, which seems a strange omission. In India, Brahmegupta (A.D. 628) gave $(a + b)^3$ in his *Arithmetic* [6].

It is to Persia, however, that the European thread can be traced back. The *Al-bahir* of Al-Samawal (died about 1180) is reported [7] as containing a calculation of the coefficients, resulting in the Binomial Triangle, which had been discovered by Al-Karaji some time soon after 1007. It is possible that Al-Karaji was inspired to make his discovery by hearing of Brahmegupta's result for the cube of a binomial, for it is believed that Brahmegupta's work had been brought to Baghdad in the eighth century, and Al-Karaji, who worked in Baghdad, drew much else from Hindu sources [8].

Al-Kashi [9], who died in Samarkand in 1429, gave, in his *Key of Arithmetic*, the Binomial Triangle up to the ninth power, together with its rule of construction (equation 5.1), but he also gave the general rule (5.2). Moreover, he was building on the work of Nasir al-Din al-Tusi who gave these methods in 1265. In the light of these facts, Omar Khayyam's oft-quoted claim [10] from about 1100 to have written a work describing the raising of a binomial to the fourth, fifth, sixth, and higher powers need not be doubted.

The binomial theorem for positive integral index can therefore be attributed to Al-Kashi (1427) at the very latest. As we noted in Chapter 4, Cardano essentially gave it in 1570 for the first time in the West. It used to be attributed to Newton, and unfortunately this error is perpetuated in Rouse Ball's popular *History* [11]: Smith [12] gives it to Pascal, but, as we shall see in the next chapter, it is the *proof* which is Pascal's contribution.

The natural format for the Binomial Triangle is symmetrical (Fig. 19), with the coefficients of $(a + b)^n$ following those of $(a + b)^{n-1}$, the addition rule of construction (5.1) being obvious from the identity

$$(a + b)^n \equiv (a + b)(a + b)^{n-1}. \tag{5.3}$$

In this form it appears on the title page of Apianus's *Arithmetic* of 1527 [13], and in the work of Tartaglia [14] (Fig. 20), who invented both the Triangle and its addition rule afresh, having already given a version of the Arithmetical Triangle in a combinatorial context [15].

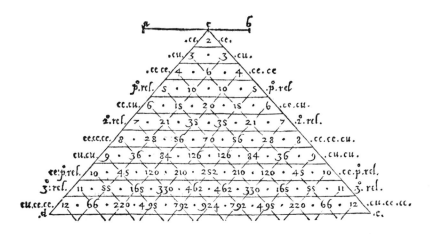

Fig. 20   The Binomial Triangle from Tartaglia's *General trattato* (1556)

It is interesting that he explicitly started from Euclid's result for $n = 2$. As we noted in Chapter 1, Stifel's version of the Triangle was really figurate in origin, whilst Scheubelius's (1545) [16] was more binomial. Instructing his readers first to place two series of natural numbers so that they meet in an apex in the number 2 (1 having been omitted), they are then to fill up the space in between with numbers formed by the addition of the two numbers next above them. Though this description suffices for us, it might have left Scheubelius's contemporaries a little bewildered had he not given a table (which clearly left its compositor *very* bewildered!): a slanted Binomial Triangle with the border of 1's missing (Fig. 21). Like Stifel's, the table was to be used for the extraction of roots. Vieta (1591) [17] did not give the Triangle itself, but obtained the coefficients up to $n = 6$ by using (5.3).

An early way to think of the consequence of (5.3) was to note that the first few rows of binomial coefficients were given by the powers of

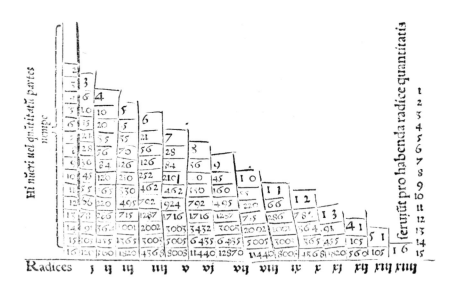

Fig. 21   The Binomial Triangle from Scheubelius's *De Numeris* (1545)

the number 11: 121, 1331, 14641. In the next row the pattern breaks down, of course, because two-digit coefficients start appearing, but this can be allowed for, and the idea was prevalent in the sixteenth century (Alexander, 1524; Forcadel, 1556) [18]. Faulhaber (1631) [19] mentions the method, and it is interesting to note that Newton thought along these lines in 1676 [20]. The explanation that the sum of the binomial coefficients is 2^n because this is $(1 + 1)^n$ was given by Wallis in 1685 [21] and W. Jones in 1706 [22].

In England the Binomial Triangle was first published by Oughtred in 1631 [23], although, as we noted in Chapter 1, Briggs probably knew both it and the binomial theorem for positive integral index (5.2) in about 1600: Hutton [24] wrote "This theorem then being thus plainly taught by Briggs about the year 1600". Napier, in Scotland, gave the Binomial Triangle in a manuscript dated about 1594 [25] but not published until 1839, in connection with the binomial expansion, complete with the addition rule. He applied it to the extraction of roots, as had Scheubelius. Harriot's unpublished manuscript from about 1610 contains the Binomial Triangle and equation (5.4) [26].

From about 1600 it becomes increasingly artificial to maintain any distinction between the Figurate, Combinatorial, and Binomial Triangles, but it is interesting to find that Hérigone (1634) fails to relate the combinatorial numbers, which he gives [27] according to the Hindu rule (our equation 3.1), to the binomial numbers, for which he gives [28] a Triangle and the addition rule. He did, however, give [29] the expansion of the trinomials $(a + b + c)^2$ and $(a + b + c)^3$, as did Oughtred [30], but the general multinomial expansion did not appear until the end of the century [31]. The complete relation

$$\binom{n}{r} = {}^nC_r = f_r^{n-r+1},\tag{5.4}$$

which is obtained from (1.9) and (4.3), is really Cardano's, as we saw in Chapter 4, but it is to Pascal that we have to turn for an extensive exploitation of this relation couched in recognizably modern terms, and for the first proper proof of the binomial theorem for positive integral index. For the extension of the theorem to non-integral and negative indices we have, of course, to await Isaac Newton (Chapter 8).

Notes to Chapter 5

[1] Sarton (1958), p. 36.
[2] Ho Peng-Yoke (1976), p. 540; Needham (1959), pp. 133–7: a manuscript copy of Yang Hui's figure dated 1407 is reproduced on page 136. Ho Peng-Yoke (1976) gives 1050 as the date at which Chia Hsien flourished, whilst Needham gives 1100.
[3] Ho Peng-Yoke (1971), p. 268; Needham (1959), pp. 133–7; Chu Shih-chieh's figure is reproduced on page 135; Lam Lay Yong (1970) gives a translation.
[4] Lam Lay Young (1980), in particular, stresses the role of the Binomial Triangle; see also Wang and Needham (1955).
[5] Smith (1925), p. 507.
[6] Brahmegupta (628), p. 279.
[7] Anbouba (1975), p. 93; Rashed (1973), p. 243.
[8] Smith (1923), pp. 167 and 283.
[9] Youschkevitch and Rosenfeld (1973), p. 257.
[10] Smith (1925), pp. 507–8.

[11] Rouse Ball (1908), p. 231.
[12] Smith (1925), pp. 510–11.
[13] Smith (1925), p. 509.
[14] Tartaglia (1556), Part II, foglio 69r–73v. In Italy the Triangle is still known as "Tartaglia's Triangle".
[15] Tartaglia (1556), Part II, foglio 17r: our Fig. 14.
[16] Scheubelius (1545), *Tractatus quintus* (pages not numbered).
[17] Vieta (1591), pp. 16–19, in *Ad Logisticen Speciosam*.
[18] Tropfke (1924), pp. 35–7.
[19] Faulhaber (1631), (pages not numbered).
[20] See Chapter 8.
[21] Wallis (1685b), p. 113.
[22] Jones (1706), p. 215.
[23] Oughtred (1631), p. 26.
[24] Hutton (1785), p. 76.
[25] Napier (1594), p. 50.
[26] See our Chapter 1.
[27] Hérigone (1634), *Arithmétique practique*, pp. 119–21.
[28] Hérigone (1634), *Algèbre*, pp. 16–19.
[29] Hérigone (1634), *Algèbre*, p. 19.
[30] Oughtred (1652), p. 62; not in the first edition (1631).
[31] See Chapter 9.

6

Pascal's Treatise on the Arithmetical Triangle, Part I

In the preceding chapters we have traced the component parts of the theory of the Arithmetical Triangle from the earliest times up to 1654, when Pascal is known to have written the *Traité du triangle arithmétique* [1].

Blaise Pascal was born on 19 June 1623 in Clermont [2]. In November 1631 his father Etienne Pascal (himself an able mathematician who gave his name to the "limaçon of Pascal") moved to Paris in order to secure his son a better education. In 1635 Etienne was one of the founders of Marin Mersenne's "Academy", the finest exchange of mathematical information in Europe at the time. To this informal Academy he introduced his son at the age of fourteen, and Blaise immediately put his new source of knowledge to good use, producing (at the age of sixteen) his famous *Essay pour les coniques* based on the work of Desargues, another founded-member of the Academy.

Thus the young and impressionable Pascal was introduced to the circle of mathematicians just as Mersenne's books containing "Pascal's Triangle" and its combinatorial applications were coming from the presses (see our Chapter 4), and he continued to visit Mersenne even when living out of Paris [3]. Moreover, Etienne Pascal was particularly noted for his knowledge of music, to which Mersenne paid tribute in both the *Harmonicorum libri XII* and the *Harmonie universelle*, having already dedicated his treatise on the organ to him [4]. It is unthinkable that these works were not in Etienne's house from the moment they were published, and we need look no further for the young Pascal's principal source. Yet there is a tendency amongst historians of science to view Pascal's *Traité* as a work of inspiration rather than consolidation [5].

Pascal was, as we shall see, a little forgetful about his sources. Practically everything in the *Traité* except the solution to the important "Problem of Points" will have been known to Mersenne's circle by 1637. It seems likely that Pascal absorbed most of this as a young man, and then, more than a decade later, his correspondence with Fermat stimulated him to compose the *Traité*, which he did in the space of a few weeks. The evidence is that, with the passage of time, he had lost most of the details whilst retaining the outline. Just as a lecturer often lectures best when, after careful preparation, he forgets his lecture notes, so Pascal poured forth his mature view of the Arithmetical Triangle and its uses, uncluttered with peripheral detail. His novel theme was to view the properties of the Arithmetical Triangle as *pure mathematics*, to be demonstrated from the fundamental addition relation independently of any binomial or combinatorial application.

Pascal probably wrote his book in August 1654. The stimulus for his interest was combinatorial, for he had succeeded in solving the gambler's Problem of Points, which had taxed Pacioli, Tartaglia and Cardano. Briefly, this problem concerns the division of the stakes between two players when a game has to be left unfinished. For example, suppose the game consists of tossing a penny, player *A* to count heads as points in his favour, and player *B* to count tails, the winner to be the first to score an agreed number of points. How should the stakes be split if the game is interrupted when *A* still needs *a* points to win, and *B* still needs *b*? It is supposed that the coin is a fair one, and that at the outset each player contributed an equal stake.

The story of how Pascal encountered the problem, and of his subsequent correspondence with Fermat about it, has been told so many times that we refer the reader elsewhere [6]. The correspondence itself, from the summer of 1654, is available in English translation [7]. We return to a consideration of Pascal's final solution in the next chapter.

The *Traité du triangle arithmétique, avec quelques autres petits traitez sur la mesme matière*, published in 1665, is made up of four separate tracts bound together (as may be seen from the pagination). The printed sheets were found amongst Pascal's papers after his death; a title-page, *Avertissement*, and list of contents were added, together with a figure of the Triangle drawn from a draft of Pascal's (see our Frontispiece), and the whole then bound and sold.

The first tract is of 11 printed pages and is entitled *Traitté du triangle arithmétique*; after the definition of the Triangle it contains nineteen corollaries and a problem. The second tract is of 8 pages, with the short title *Divers usages du triangle arithmétique*, and has two parts: *Usage du triangle arithmétique pour les ordres numériques* and *Usage du triangle arithmétique pour les combinaisons*. The third tract is of 16 pages and also consists of two parts: *Usage du triangle arithmétique pour déterminer les partys qu'on doit faire entre deux joueurs qui jouent en plusieurs parties* and *Usage du triangle arithmétique pour trouver les puissances des Binomes et Apotomes*. The fourth tract starts with the *Traitté des orderes numériques*, which is in French, with eleven propositions, and is followed by six parts in Latin. It has 48 pages.

The question arises as to how much of this material might properly be regarded as comprising the *Traité du triangle arithmétique* itself. The pagination, the typography of the section headings, and the contents themselves, all point to slightly different conclusions, but it seems reasonable to regard the title of the second tract, *Divers usages du triangle arithmétique*, which is on a separate title-page, as encompassing the two sections of the third tract as well, even though the pagination starts again, thus leaving us with a book of 36 pages (one blank) in two parts, whose English description might be:

A Treatise on the Arithmetical Triangle
by B. Pascal

Part I   *A Treatise on the Arithmetical Triangle*
Part II *Uses of the Arithmetical Triangle*
      (1) In the theory of figurate numbers.
      (2) In the theory of combinations.
      (3) In dividing the stake in games of chance.
      (4) In finding the powers of binomial expressions.

Part I and Part II Sections 1 and 2 at least must have been completed before 29 August 1654 because on that date Fermat [8] wrote to Pascal referring to the *Traités du Triangle arithmétique et de son application*, but that reference does not necessarily include Part II Sections 3 and 4. Sections 3 and 4 are probably later than 29 July, for Section 3 goes further than Pascal's letter of 29 July [9] in applying the theory of combinations to the Problem of Points.

From the above description it will be seen that Pascal's book encompasses the three main topics we have treated so far, the figurate numbers, the combinatorial numbers, and the binomial numbers. The remaining *Traitté des ordres numériques* and its associated parts do not add very much of interest on these topics; most of their propositions are extensions of those of the *Traité du triangle arithmétique*. Mesnard [10] describes their relation to the rest of the work; we discuss them in the next chapter.

There appears to be no complete English translation of the *Treatise on the Arithmetical Triangle* as above described, but Smith [11] gives a translation of nearly the whole of Part I and of Part II Section 4. I am not aware of any extensive description of the *Treatise* using modern notation and terminology other than Bosmans's account of Part I published in 1923 [12]. What follows here is not so much a description of the work but an extraction of its results, with a commentary on their origins.

Part I consists, as we have noted, of the definition of the Triangle, nineteen corollaries, and a problem. The definition is in terms of the addition formula

$$f_k^l = f_k^{l-1} + f_{k-1}^l. \qquad (1.6 \ bis)$$

(We continue using the notation for the figurate numbers established in Chapter 1; Pascal did not use a general notation, as may be seen from our Frontispiece. The notation f_k^l has the disadvantage that since it refers to the lth figurate number in k dimensions (Fig. 8) it becomes the number in the lth column and $(k + 1)$th row of Pascal's Triangle, but it seems best not to introduce yet another notation; it is, however, convenient to recall and to use

$$\binom{n}{r} = {}^nC_r = f_r^{n-r+1} \quad \text{as well.}) \qquad (5.4 \ bis)$$

Pascal allows the corner number f_0^1, which he calls the *generator* of the triangle, to be any integer and not just 1, but this generalization is of no interest and we set $f_0^1 = 1$.

Corollary 1

$$f_k^1 = f_0^l = f_0^1 = 1; \quad l = 2, 3, 4, \ldots; \quad k = 1, 2, 3, \ldots. \quad (6.1)$$

Compare our equation (1.6); Pascal regards this as a corollary rather than as part of the definition, arguing as if a row and a column of zeros existed outside the triangle.

Corollary 2

$$f_k^l = \sum_{i=1}^{l} f_{k-1}^i. \quad (6.2)$$

This is simply our (1.7).

Corollary 3

$$f_k^l = \sum_{j=0}^{k} f_j^{l-1}. \quad (6.3)$$

This is the same proposition, but summing columns instead of rows. Pascal proves both these corollaries by successive substitution.

Corollary 4

$$f_k^l = \sum_{i=1}^{l-1} \sum_{j=0}^{k-1} f_j^i + 1. \quad (6.4)$$

In words, the sum of all the numbers enclosed by the row and column of a number, plus 1, equals that number. Pascal demonstrates this corollary by successive substitution using Corollaries 2 and 3.

Corollary 5

$$f_k^l = f_{l-1}^{k+1}, \quad \text{or} \quad \binom{n}{r} = \binom{n}{n-r}. \quad (6.5)$$

This is the recognition of the symmetry of the whole Triangle about the main diagonal; Pascal notes that the numbers in the *base* of each triangle (that is, the numbers $\binom{n}{r}$ or f_r^{n-r+1}, $r = 0, 1, 2, \ldots, n$, for fixed n) are symmetrically derived from the numbers in the base of the preceding triangle $(n - 1)$, and so on down to $f_0^2 = f_1^1$, which is true by Corollary 1. Such symmetry was well known, though possibly not before Stifel for the figurate numbers considered in isolation (see Chapter 1).

Corollary 6
Corollary 6 simply states that corresponding numbers in a row and its symmetrically placed column are the same, and is thus a special case of Corollary 5. Bosmans's equation [13], though not incorrect, is not what Pascal actually meant.

Corollary 7

$$\sum_{r=0}^{n} \binom{n}{r} = 2 \sum_{r=0}^{n-1} \binom{n-1}{r}. \qquad (6.7)$$

Each number in base $(n - 1)$, argues Pascal, is represented twice in base n by virtue of the fundamental definition.

Corollary 8

$$\sum_{r=0}^{n} \binom{n}{r} = 2^n. \qquad (6.8)$$

In consequence of Corollary 7; known to Cardano in 1570 (see equation 4.5).

Corollary 9
The sum of the numbers in a base is equal to the sum of the sums of the numbers in the lower bases, plus 1; in other words, using Corollary 8,

$$2^n = 2^{n-1} + 2^{n-2} + \ldots + 2 + 1 + 1, \qquad (6.9)$$

"because this is a property of the 'doubling' progression", says Pascal.

Corollary 10

$$\sum_{r=0}^{s} \binom{n}{r} = \sum_{r=0}^{s} \binom{n-1}{r} + \sum_{r=0}^{s-1} \binom{n-1}{r}. \qquad (6.10)$$

Another direct consequence of the fundamental definition. Savitsky's translation [14] of this corollary is poor, and Bosmans's equation [15] is not what Pascal actually proposed, though it comes to the same thing. An example shows the structure of this corollary most clearly:

$$
\begin{array}{rrr|rrr}
1 & 4 & 6 & 4 & 1 & \\
 & 1 & 4 & 6 & 4 & 1 \\
\hline
1 & 5 & 10 & 10 & 5 & 1 \\
\end{array}
$$

It is, of course, nothing more than the familiar trick of "multiplying by 11" mentioned in the last chapter.

Corollary 11

$$f_k^{k+1} = 2f_k^k = 2f_{k-1}^{k+1}, \quad \text{or} \quad \binom{2n}{n} = 2\binom{2n-1}{n-1} = 2\binom{2n-1}{n},$$

$$(6.11)$$

an immediate result for the middle number of a base.

Pascal continues:

"Note: All these corollaries are on the subject of the equalities which one meets in the Arithmetical Triangle. Now we shall take a look at the proportions, for which the following proposition is the foundation".

Corollary 12

"Of two adjacent coefficients in a base, the upper is to the lower as the number of coefficients from the upper onwards is to the number of coefficients from the lower downwards". Consider the base with index n:

$$1, n, \binom{n}{2}, \binom{n}{3}, \ldots, \binom{n}{r-1}, \binom{n}{r}, \binom{n}{r+1}, \ldots, \binom{n}{n-3}, \binom{n}{n-2}, n, 1.$$

$\underbrace{\qquad\qquad\qquad\qquad}_{r \text{ coefficients}}$ $\underbrace{\qquad\qquad\qquad\qquad\qquad}_{(n-r+1) \text{ coefficients}}$

Pascal's statement is, in the language of proportions,

$$\binom{n}{r} : \binom{n}{r-1} = (n-r+1) : r, \quad \text{or}$$

$$\binom{n}{r} = \frac{n-r+1}{r} \binom{n}{r-1}. \qquad\qquad (6.12)$$

In the notation of the figurate numbers,

$$rf_r^{n-r+1} = (n-r+1)f_{r-1}^{n-r+2},$$

or, putting $r = k$ and $n - r + 1 = l$,

$$kf_k^l = lf_{k-1}^{l+1}. \qquad\qquad (1.17 \text{ bis})$$

This, as we have noted, is Cardano's rule (our equation 4.4) from 1570, given also by Briggs in about 1600 (equation 1.17) and rediscovered by Fermat in 1636. Pascal proves this important corollary by an explicit use of mathematical induction, which method in its full generality has therefore often been attributed to him [16]:

"Although this proposition has an infinite number of cases, I will give a short demonstration, using two lemmas.

Lemma 1, which is self-evident, is that this proportion is met with in the second base.

Lemma 2 is that if this proportion is found in any base, it will necessarily be found in the next base".

The second Lemma is then easily proved by using the fundamental definition, whence the corollary follows by induction.

This is the "twelfth proposition" to which Fermat referred in his letter to Pascal dated 29 August 1654 [17], stating that it had come through the post from Paris to Toulouse (as part of the *Traité*) at just the moment when his proposition on the figurate numbers, which was the same thing, was travelling from Toulouse to Paris (equations 6.12 and 1.17 respectively). No letter of Fermat's exists conveying his proposition to Pascal, but at the end of the *Traité des ordres numériques* Pascal describes how the two authors had found the proposition independently "as our letters written and received at the same time testify". Evidently Pascal did not know that Fermat had found the result eighteen years previously, in 1636; and just as Fermat had then regarded it as "beautiful and wonderful", so Pascal rejoiced in his own discovery. But here we must introduce a note of caution. At the beginning of the chapter we suggested that Pascal was sometimes a little forgetful about his sources. Even if he did not encounter Fermat's result at a meeting of Mersenne's Academy (Fermat having informed both Mersenne and Roberval, a member, in 1636 [18]), he is likely to have seen the rule in *Harmonicorum libri XII* (1636) and *Harmonie universelle* (1637), where Mersenne used it to derive the combinatorial coefficients for $n = 36$, $r = 1, 2, 3, \ldots$ (Chapter 4). Though Pascal's strength was in plundering the Figurate Triangle *and* the Combinatorial Triangle *and* the Binomial Triangle for relations amongst their numbers, he was not the first to identify these Triangles: Corollary 12 may have seemed original in 1654 when given out of its combinatorial or figurate context, but there is more than a suspicion that it must have lodged in his mind in the later 1630's. It is very easy to "discover" something you knew fifteen years previously [19].

Corollary 13

$$f_k^l = \frac{l + k - 1}{k} f_{k-1}^l.$$
(6.13)

Used by Briggs (cf. equation 1.19 and its derivation).

Corollary 14

$$f_k^l = \frac{l + k - 1}{l - 1} f_k^{l-1}.$$
(6.14)

The equivalent of Corollary 13 by symmetry, as may be seen by using Corollary 5 and relabelling the rows and columns.

Corollary 15

$$\sum_{i=1}^{l} f_k^i = \frac{l+k}{k+1} f_k^l.$$
(6.15)

This, as Pascal observes, is a direct consequence of Corollaries 2 and 13.

Corollary 16

$$\sum_{i=1}^{l} f_k^i = \frac{k+2}{l-1} \sum_{i=1}^{l-1} f_{k+1}^i.$$
(6.16)

By Corollaries 2 and 12; Pascal gives two verbal versions.

Corollary 17

$$\sum_{j=0}^{k} f_j^l = \frac{k+1}{l} \sum_{i=1}^{l} f_k^i.$$
(6.17)

By Corollaries 2, 3 and 12.

Corollary 18

$$\sum_{i=1}^{l} f_k^i = \frac{l}{k+1} \sum_{i=1}^{k+1} f_{k-1}^k.$$
(6.18)

By Corollaries 6 and 17; again Pascal gives two verbal versions.

Corollary 19 (and last)

$$f_k^{k+1} = 4 \frac{2k-1}{2k} f_{k-1}^k.$$
(6.19)

By Corollaries 11 and 14.

The culmination of Part I is the solution to the following:

Problem

Given l and $k + 1$, find f_k^l.

Solution

By Corollary 12,

$$kf_k^l = lf_{k-1}^{l+1} \qquad (1.17 \ bis)$$

so that

$$f_{k-1}^{l+1} = \frac{k}{l} f_k^l$$

and

$$f_k^l = \frac{k+1}{l-1} f_{k+1}^{l-1},$$

which, recursively applied, gives

$$f_k^l = \frac{(k+1)(k+2)(k+3)\dots(k+l-1)}{(l-1)(l-2)(l-3)\dots 1}. \qquad (6.20)$$

Had Pascal proceeded recursively to the other end of the base he would have found the equivalent result, dictated by symmetry, which we may obtain formally by applying Corollary 5 and thus replacing $(k + 1)$ by l and $(l - 1)$ by k on the right-hand side of (6.20) to obtain

$$f_k^l = \frac{l(l+1)(l+2)\dots(l+k-1)}{k(k-1)(k-2)\dots 1}. \qquad (1.18 \ bis)$$

This result had been known for some time, having been given by Cardano in 1570 as we have seen.

Summarizing Part I of the *Traité*, we may say that it contains, first, a clear formal account of eleven simple propositions arising from the fundamental addition rule for generating the numbers of the Triangle, most of which were already known (Corollary 4, which is pleasing, may have been new), and, secondly, an explicit proof by induction of the multiplicative rule for deriving each number from its neighbour in the same base (Corollary 12), followed by seven consequential propositions culminating in the general formula. Neither the multiplicative rule itself nor the resultant general formula were new, though each has often been attributed to Pascal. Since (as we see in the next chapter) Pascal included amongst the uses of the Arithmetical Triangle the expansion of a binomial with positive integral index, his proof of the general formula constitutes a proof of the binomial theorem for positive integral index. This is often attributed to James Bernoulli's *Ars conjectandi* of 1713 [20].

Notes to Chapter 6

[1] Pascal (1665). Two modern French editions of Pascal which include the *Traité* are those of L. Lafuma from 1963 (Pascal 1654a, pp. 50–94) and J. Mesnard from 1964 to 1970 (Pascal, 1654b, Volume II, pp. 1176–1332). The latter may be used for preference if the original is not available; the former unaccountably interchanges the last two sections of the *Traité des ordres numériques*. Mesnard gives a valuable introduction (Vol. II, pp. 1166–75), as well as the text of a Latin version (Vol. II, pp. 1196–1214) which he believes to be earlier than the French.

[2] For information on Pascal's life see the *Oeuvres* edited by Mesnard (Pascal 1654b, Vol. II) and the numerous biographies. Readers who wish for something in English may note especially Mesnard (1952) and Mortimer (1959). Bosmans (1924) and Boyer (1963) have concentrated on Pascal's mathematical work.

[3] Pascal (1654b), Vol. II, p. 237.

[4] Pascal (1654b), Vol. II, pp. 119–22.

[5] Thus Sarton (1958), p. 37: "... many people believe that Pascal's triangle was an original discovery without antecedents. The discovery may have been original so far as Pascal was concerned, but it had moved capriciously from mind to mind; it had been 'in the air' for centuries"; Boyer (1968), p. 398: "The triangle itself was more than 600 years old, but Pascal disclosed some new properties" [Boyer then quotes Corollary 12 which, as we see later in the chapter, was not only known, but actually given in the book from which Pascal almost certainly learnt of the Triangle]; David (1962), pp. 81–2, is immune from this criticism.

[6] Edwards (1982b), reprinted in the present volume, Appendix I; some earlier accounts are Todhunter (1865), pp. 7–16; Ore (1960); David (1962), pp. 70–97; Mahoney (1973), pp. 390–8; Maistrov (1974), pp. 39–48; Hacking (1975), pp. 49–62; and Sheynin (1977), pp. 231–9. Note that the date Maistrov gives for the publication of the *Traité* is wrong. Another account is to be found in the Introduction to Volume XIV of the works of Huygens (Huygens, 1655, pp. 21–5), whilst Hara (1962) contains much relevant discussion.

[7] Smith (1929), pp. 546–65 (translation by V. Sanford); David (1962), pp. 229–53 (translation by M. Merrington). For the French originals see Fermat

(1636), Vol. II, pp. 288–314 or Pascal (1654b), Vol. II, pp. 1131–58. Edwards (1982b), p. 260, corrects an important error common to both the above translations; see note 8 below for further corrections.

[8] David (1962), p. 82, says that Pascal mentioned the Arithmetical Triangle in a letter to Fermat in August, but this seems to be a mistake; Sheynin (1977), p. 234, confirms that "in neither of his letters does Pascal refer to the arithmetical triangle". The apparent reference to Pascal's "eleventh proposition" in Fermat's letter has caused a great deal of trouble, but Mesnard (Pascal, 1654b, Vol. II, p. 1154 footnote) has been back to the original manuscript and recovered the correct rendering—"twelfth proposition"; all earlier editions are incorrect, and the following footnotes can be disregarded: Fermat (1636), Volume II, p. 308; Smith (1929), p. 561; and David (1962), p. 94. David's footnote is particularly confusing because she actually quotes from the eleventh proposition of the *Traité des ordres numériques* where Pascal gives a French version of Fermat's Latin proposition from 1636, though she refers to it as the "eleventh proposition on the Arithmetic Triangle", and her English translation of Fermat's proposition has *added to* where it should have *multiplied by*, thus producing nonsense. The error is, however, excusable linguistically because Pascal had used the word *mener*, and apparently it is only here and in his correspondence with Fermat that *mener* has this sense; other seventeenth-century French writers used *multiplier*. I am indebted to Professor Nais of the University of Nancy for this information, and to Mrs E. M. Harding for obtaining it for me.

[9] Pascal (1654b), Vol. II, p. 1134.

[10] Pascal (1654b), Vol. II, pp. 1166–75.

[11] Smith (1929), pp. 67–79 (translated by A. Savitsky); a lesser part is translated in Struik (1969), pp. 21–6.

[12] Bosmans (1923).

[13] Bosmans (1923), p. 459.

[14] Smith (1929), p. 72.

[15] Bosmans (1923), p. 460.

[16] See Chapter 7, note 13.

[17] See note 8 above.

[18] See Chapter 1.

[19] Bosmans (1924), p. 28, described it as "l'un des plus beaux fleurons de la couronne mathématique de Pascal"!

[20] Hutton (1786), pp. 65–82, considers the history of the theorem and says "One of the first demonstraters of this theorem, was Mr James Bernoulli"; Todhunter (1865), p. 65, says "James Bernoulli gives . . . a demonstration of the Binomial Theorem for the case of a positive integral exponent. Maseres considers this to be the first demonstration that appeared"; more recently Boyer (1968), p. 458, says "[In *Ars conjectandi*] we find the first adequate proof of the binomial theorem for positive integral powers". Boyer adds that James Bernoulli gave Pascal credit for the binomial theorem with general exponent "but this attribution appears to be gratuitous"; however, Boyer is here wrong twice over, for it was *John* Bernoulli who was alleged by Hutton (1785), pp. 41–121, to have attributed the general theorem to Pascal, but this was an error of translation by Hutton, for John Bernoulli (1742), p. 173, only attributed to Pascal a method of proof, namely mathematical induction. We describe *Ars conjectandi* in Chapter 10.

7

Pascal's Treatise, Part II, and associated tracts

Section 1 of Part II of the *Traité* applies the Arithmetical Triangle to the figurate numbers. Pascal names the series of figurate numbers in each dimension *ordres numériques*, but uses the familiar names *triangulaires, pyramidaux*, and *triangulo-triangulaires* for f_2^1, f_3^1 and f_4^1 just as Vieta and Fermat had done. This Section is very short, and simply points out the identity of the figurate numbers with the numbers of the Arithmetical Triangle, whose properties therefore apply to them. There is nothing new, but at the end Pascal promises us a separate little treatise, no doubt the *Traité des ordres numériques*.

Section 2 applies the Arithmetical Triangle to the theory of combinations. After some preliminary remarks we come to the important Lemma 4: the number of combinations of $(n + 1)$ things taken $(r + 1)$ at a time is equal to the sum of the number of combinations of n things taken r at a time and the number of combinations of n things taken $(r + 1)$ at a time, or

$$^{n+1}C_{r+1} = {}^nC_r + {}^nC_{r+1}. \tag{7.1}$$

For consider any particular one of the $(n + 1)$ things: ^nC_r gives the number of combinations that contain it, whilst $^nC_{r+1}$ gives the number that exclude it, the two numbers together giving the total. Pascal does not give a formal proof by induction but uses the above argument with $n = 3$ and $r = 1$, indicating its generality [1]. He then applies the Lemma in the demonstration that these combinatorial numbers are the numbers in the Arithmetical Triangle,

$$^nC_r = f_r^{n-r+1}, \tag{4.3 bis}$$

for, of course, (7.1) is but the fundamental addition formula of the Triangle, and the initial conditions correspond as well. Possibly he was inspired to think of (7.1) by Hérigone's 1634 result [2] that of the ^nC_r combinations of n things taken r at a time, $^{n-1}C_{r-1}$ will include any particular thing, from which (7.1) follows (with n and r each increased by 1) since we already know that the number of combinations *not* including it is simply $^{n-1}C_r$. This Section also ends with the promise of a further little treatise on the matter, presumably that part of the *Traité des ordres numériques* entitled *Combinationes*, which indeed opens with a repeat of Section 2.

Though, as we have seen, the identification of the combinatorial numbers with the figurate numbers goes back to Cardano, Pascal has here made a major contribution by finding the direct combinatorial proof of the fundamental addition formula. Later writers have often either overlooked this (for example, Todhunter [3]) or have gone too far in the other direction by attributing the identification itself to Pascal (for example, Montmort [4]). Indeed it is to Montmort that Pascal probably owes his reputation over the Arithmetical Triangle, for his textbook *Essay d'analyse sur les jeux de hazard*, whose principal edition is dated 1713, starts with a *Traité des combinaisons* which is heavily dependent on Pascal's *Traité*. Moreover, in his Introduction [5] Montmort wrote that "... Pascal has proceeded furthest, as is clear from his treatise *The Arithmetical Triangle*, which is full of observations and discoveries on the figurate numbers of which I believe him to be the originator, since he does not cite any other person". Of the "beautiful theorem" (Corollary 12) Montmort does remark that Pascal associates Fermat with the discovery, but by not being aware that in its combinatorial context it was published by Cardano in 1570, he reinforces the impression that it was first found by Pascal and Fermat.

Section 3 of Part II applies the Arithmetical Triangle to the Problem of Points. As we remarked in the last chapter when describing the problem, the history of Pascal's involvement in it has so often been told that we refer the reader elsewhere [6]. Here we shall only discuss it in enough detail to be able to comment on Pascal's application of the Arithmetical Triangle to it.

Suppose that in order to win, A needs a points and B needs b points. Both Pascal and Fermat solved the problem initially by the method of "combinations", as they called it. This involved the realization that at most $(a + b - 1)$ more tosses will settle the game, and that if this number of tosses is imagined to have been made, the resulting 2^{a+b-1} possible games, each equally probable, may be classified into those which A wins and those which B wins, the stakes then being divided in this proportion. Thus the real game, of indeterminate length, is embedded in an imaginary game of fixed length [7]. Pascal gave an example of this procedure in his letter of 24 August 1654, taking $a = 2$ and $b = 3$; at most four throws will be necessary, giving 16 imaginary games. These are enumerated, and Pascal notes that "because the first player needs two wins, he must win the game whenever there are two [heads]: thus there are 11 [games] for him; and because the second player needs three, he must win the game whenever there are three [tails]: therefore there are 5 [games] for him".

Now the remarkable feature of this solution is that the order of occurrence of the heads and tails does not matter: only the total numbers of each. The reason is that, in the imaginary game of $(a + b - 1)$ tosses, even if A only scores the a heads he needs to win, and no more, B, playing on to the bitter end, only scores $(b - 1)$ tails, one short. And if A scores more than a heads, B is even further away from winning. In other words, if the imaginary game contains $0, 1, 2, \ldots, (a - 1)$ heads *in any order*, B will have won the game, whilst if it contains $a, (a + 1), (a + 2), \ldots, (a + b - 1)$ heads A will have won the game. Playing on so as to complete $(a + b - 1)$ tosses makes no difference. The representation by lattice diagram (Fig. 22 and 23) makes this clear. As soon as three players are involved, using a three-sided die, the order of occurrence of the throws *does* matter in general, a fact which led to some discussion between Pascal and Fermat.

We must now turn aside from the combinatorial solution for a moment to introduce Pascal's alternative method, given in his letter of 29 July 1654. In modern terms, this involved analysing the tree of possible real games (Fig. 24) and, working backwards from the tips, using recursively the notion that, if expectations of gain of X and Y units are equally probable, the expectation of gain is $\frac{1}{2}(X + Y)$ units [8]. Furthermore, the stakes are to be divided according to the expectation of gain, that is, the value of a gamble is equal to its expectation. We shall refer to this method as "Pascal's", and to the solution by enumeration as "Fermat's", as they did themselves during the correspondence [9].

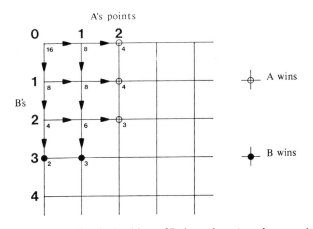

Fig. 22   Lattice diagram for the Problem of Points when A needs two points to win and B three. A step to the right represents a point for A, and a step downwards a point for B. Starting with 16 games, the numbers in the lattice show the expected number of games at each stage, ending in 11 wins for A and 5 for B

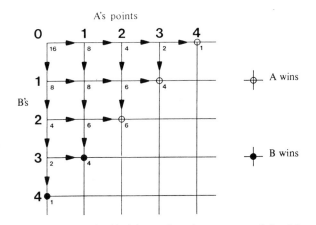

Fig. 23   The real game embedded in an imaginary game of fixed length, four throws. The numbers in the lattice show the expected number of games at each stage, ending in 11 wins for A and 5 for B. It is clear, by comparing this with Fig. 22, that A wins if and only if he would have won the original game, and similarly for B

Pascal was unable, however, to generalize his method fully at the time of the correspondence, although he did manage to solve the case where A needs but one win ($a = 1$) whilst B still needs many (b): as we should now argue, in order for B to win the game he must win each of the next b points, an event with probability $(1/2)^b$, which is therefore the proportion of the total stake he should receive in

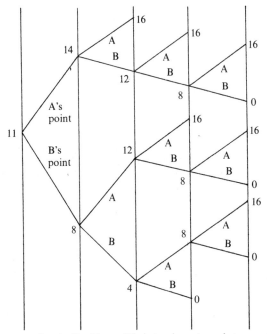

Fig. 24　Event-tree for the Problem of Points when *A* needs two points to win, and *B* three. A stake of 16 is available to the winner. The tree shows, at its tips, *A*'s winnings, and, at its nodes, *A*'s expected winnings calculated by Pascal's argument. Thus, as the game stands, the stake should be divided 11 to *A* and 5 to *B*. (Modified from Jacquard (1974))

compensation for giving up the unequal struggle! Pascal saw this solution inductively, as we can by drawing an event-tree.

It was when Pascal attempted, in his letter, to resolve another partially general case that the Arithmetical Triangle made its appearance: he admitted to being unable to use his method to solve the case in which *A* has just one point, and *B* none, so that if *B* wants *b* points, *A* wants (*b* − 1). An attempt to draw the corresponding event-tree will show why Pascal's method was difficult: it is as complicated as in the completely general case. But he did manage to solve it using Fermat's method assisted by a short cut involving the combinatorial numbers. Embedding the real game in an imaginary game of (2*b* − 2) points, as described above, he must have understood that if there are 0, 1, 2, . . . , (*b* − 2) heads *B* will win, but if (*b* − 1), *b*, (*b* + 1), . . . , (2*b* − 2) *A* will win. We now put *m* = *b* − 1 to ease the notation. Pascal is then able to sort the 2^{2m} games into these classes and to find the corresponding numbers in each, $^{2m}C_0$, $^{2m}C_1, \ldots, {}^{2m}C_{m-1}$ for the cases where B wins, and $^{2m}C_m$, $^{2m}C_{m+1}$, $\ldots, {}^{2m}C_{2m}$ for the cases where A wins, because he uses the theorem

$$\tfrac{1}{2} \cdot {}^{2m}C_m + {}^{2m}C_{m+1} + \ldots + {}^{2m}C_{2m} = 2^{2m-1}, \tag{7.2}$$

obviously with this in mind. He gives no proof of the theorem, but presumably he saw it as an immediate consequence of the symmetry of the Arithmetical Triangle which he was then in the process of investigating (Corollaries 5 and 8 together give it). Thus of the 2^{2m} equiprobable games, A wins

$$2^{2m-1} + \tfrac{1}{2} \cdot {}^{2m}C_m, \tag{7.3}$$

found by adding the missing half-coefficient to (7.2). If, therefore, the players had staked equal amounts, A should receive not only his own stake back, but $\tfrac{1}{2} \cdot {}^{2m}C_m/2^{2m}$ of the total as well, which is $^{2m}C_m/2^{2m}$ of B's.

Pascal does not explain his method so fully nor so generally, since he uses the specific example $b = 5$ (and hence $m = 4$, $2m = 8$): "Thus, if I have won the first point out of 5, $35/128$ of my opponent's stake is due to me": for $^8C_4 = 70$ and $2^8 = 256$. Moreover, he takes the calculation one stage further by using the relation

$$^{2m}C_m/2^{2m} = \frac{1 \cdot 3 \cdot 5 \ldots (2m-1)}{2 \cdot 4 \cdot 6 \ldots 2m}. \tag{7.4}$$

How did he know this? He gives no clue ("You will undoubtedly understand all this well, if you take a little trouble", he writes to Fermat; and we must do so too). But we have the advantage of hindsight, for, having studied *The Arithmetical Triangle* before reading the correspondence with Fermat, we recognize (7.4) as Corollary 19, which, replacing k by m and using the combinatorial notation, is

$$^{2m}C_m = 4\frac{2m-1}{2m}\,{}^{2m-2}C_{m-1}. \tag{6.19 bis}$$

By repeated application (7.4) is immediately obtained.

We may note as an aside that if we assume that Pascal obtained (7.4) by the same route as he proved Corollary 19, then we know that by the date of this letter (29 July 1654) he must have possessed Corollaries 5, 11, 12, 13, 14 and 19, and we have already seen the evidence that he knew Corollary 8, and therefore 7, as well (in the derivation of 7.2).

All commentators agree that Fermat must have replied to this letter of Pascal's dated 29 July 1654, probably on 9 August when he wrote to Carcavi [10], but no letter has been found. Fermat's mathematical biographer, Mahoney [11], goes so far as to attribute the above reasoning (down to equation 7.3) to the missing reply, postulating that "Though Fermat agreed with Pascal's numerical results, he did not much like the method used to obtain them". But what, then, *was* that method?

This is as much of the Fermat–Pascal correspondence as we need study in order to understand Section 3 of Part II of *The Arithmetical Triangle*. The important point is that Pascal had recognized that the numbers in a base of the Arithmetical Triangle corresponded to the numbers of games of length n with 0, 1, 2, 3, ..., n heads, and therefore that Fermat's method could be applied without exhaustive enumeration of all the possibilities. In the general case, where A wants a and B wants b points, there would be 2^{a+b-1} imaginary games,

$$\sum_{r=0}^{a-1} \binom{a+b-1}{r} \qquad \text{favourable to } B$$

and   (7.5)

$$\sum_{r=a}^{a+b-1} \binom{a+b-1}{r} \qquad \text{favourable to } A,$$

and the stakes should be divided accordingly. This is the *result* given in *Problem I—Proposition I* of Section 3, but not the proof [12]. Pascal removes the combinatorial ladder by which he surely ascended, and proves the general result by induction based on his method of expectation, described above. It is a brilliant performance, and we shall give it in its general form (though Pascal, as usual, worked through an example and then noted the generality of the argument).

Proof

Suppose the theorem (see 7.5) true, and suppose that A needs a points to win and B needs b. After the next toss either A will still need a points, having lost the toss, or he will need $a - 1$ points having won it. In the former case, B's expected proportion of the total stake will be (by 7.5)

$$\sum_{r=0}^{a-1} \binom{a+b-2}{r} \Bigg/ \sum_{r=0}^{a+b-2} \binom{a+b-2}{r} \qquad (7.6)$$

and in the latter

$$\sum_{r=0}^{a-2} \binom{a+b-2}{r} \Bigg/ \sum_{r=0}^{a+b-2} \binom{a+b-2}{r}. \qquad (7.7)$$

By Pascal's method of expectations, therefore, B's expected proportion prior to the toss will be the mean of (7.6) and (7.7), since each corresponds to an event of probability (as we should now say) one half, and this is

$$\frac{\sum_{r=0}^{a-1} \binom{a+b-2}{r} + \sum_{r=0}^{a-2} \binom{a+b-2}{r}}{2 \cdot \sum_{r=0}^{a+b-2} \binom{a+b-2}{r}}. \qquad (7.8)$$

Following Pascal precisely, we now apply Corollary 10 to the numerator and Corollary 7 to the denominator, with $n = a + b - 1$ and $s = a - 1$, whence B's expectation becomes

$$\sum_{r=0}^{a-1} \binom{a+b-1}{r} \Bigg/ \sum_{r=0}^{a+b-1} \binom{a+b-1}{r}, \qquad (7.9)$$

which is equivalent to (7.5). On the assumption, therefore, that the theorem is true when the sum of the points required by the players is $(a + b - 1)$, it is true when the sum is $(a + b)$; but it is true when the sum is 2, for then each player needs 1 point and has expectation $\frac{1}{2}$ in accordance with the theorem, and therefore, by induction, it is generally true.

Thus does Pascal crown his work on the Problem of Points with a masterly proof which uses both the idea of mathematical induction and the concept of mathematical expectation. Oddly enough, he is often wrongly credited with the invention of the former [13], whilst his notion of expectation is often attributed to Huygens in 1657 (see Chapter 10). By not revealing the origin of his solution, however, he conceals his discovery of the binomial distribution for equal chances. The evidence is overwhelming that at the time he wrote *A Treatise on the Arithmetical Triangle* Pascal knew that the numbers of (imaginary) games of length n with $0, 1, 2, \ldots, n$ points for A were given by the corresponding numbers in the Arithmetical Triangle, and thus that these numbers, divided by their sum, gave successively the *probabilities* of obtaining $0, 1, 2, 3, \ldots, n$ heads in n tosses of a fair coin. "If", writes Sheynin [14], "the tabular form of defining a set of numbers is recognized on a par with the analytical form, then, for the binomial distribution with $p = q$, Pascal's method is equivalent to the method of generating functions. For this reason I am inclined to begin the prehistory of generating functions with Pascal". There is no need to suppose that in his derivation Pascal used any argument other than the straightforward inductive one of considering each contested point in succession and noting the obvious relevance of the addition rule (see Fig. 23).

David, in her widely-read account [15] of the correspondence between Fermat and Pascal, rather takes the side of Fermat. "On the whole it would seem that Pascal's reputation as a mathematician is greater than he deserved, although there is no doubt that he was very competent". But David's account of the Problem of Points is incomplete, since it stops with the correspondence, whereas the most striking evidence of Pascal's originality is here, in the *Treatise*.

Once again, at the end of a Section, Pascal refers to a little treatise he has written, which takes the matter further and applies the corollaries of *The Arithmetical Triangle* to the division of stakes. It is likely that he was thinking of *La géométrie du hazard* mentioned in his letter to the *Académie Parisienne* [16], but in both remarks he seems to have been anticipating, and no such treatise has come to light. Should it do so, it will surely contain the binomial distribution for equal chances. Ironically, Pascal's name *is* sometimes used in connection with a statistical distribution, the *negative binomial* distribution of the number of "trials" needed to score just r

"successes", but there seems to be no direct justification for this any more than there is for "Pascal" to be an SI unit of pressure or the name of a computer language [17].

Finally, Section 4 of Part II gives a brief statement of the fact that the powers of binomial expressions (such as $(A + 1)$, $(A + 2)$, or $(A - 1)$) may be found by using the bases of the Arithmetical Triangle. His treatment reveals a clear understanding that

$$(a + b)^n = a^n + \binom{n}{1}a^{n-1}b + \binom{n}{2}a^{n-2}b^2 + \ldots + b^n, \quad (7.10)$$

which was, of course, well known. Indeed, Pascal does not offer any demonstration "because others have treated it, such as Hérigone [18]; besides, the matter is self-evident".

In sum, *A Treatise on the Arithmetical Triangle* (Parts I and II) is a clear, concise statement of existing knowledge about the Arithmetical Triangle and its uses, constructed to an original plan whereby the properties of the numbers of the triangle are first established, and then applied in various fields. In style it is modern, listing propositions, lemmas, and corollaries in an orderly fashion, and as part of this methodical approach it contains clear examples of proof by induction, and in particular the first rigorous proof of the binomial theorem for positive integral index.

Its only other contributions of substantial importance both concern combinations: the first is the direct combinatorial proof of the addition property, and the second is the implicit identification of each base of the Triangle with what we now call the binomial distribution (for $p = \frac{1}{2}$), and its use in solving the Problem of Points.

Pascal regarded his taming of the vagaries of chance by the formal application of mathematical proof as a major contribution in itself; hitherto, chance phenomena had been treated by a collection of rules, as we have seen, but Pascal took the notion of formal proof from geometry and applied it to chance, resulting in "a reconciliation of apparent opposites" [19]. "*Anceps fortuna aequitate rationis reprimitur*", he wrote "Proper calculation masters fickle fortune", so that for the Problem of Points with two players "each player always has assigned to him precisely what justice demands". The very name of his proposed work *The geometry of chance*—"this stunning title", as he called it—reflects his feelings, and it is interesting also to note the geometrical flavour of the title of the *Traité* itself: it concerns the arithmetical *triangle*.

Of the Treatises printed with *The Arithmetical Triangle*, we have already mentioned the first, the *Traitté des ordres numériques*, of which the first part is in French and the second in Latin. The French part contains eleven propositions about figurate numbers which arise through the straightforward application of the corollaries of *The Arithmetical Triangle*, including Corollary 12, in connection with which Pascal comments on Fermat's version [20]. The Latin sequel is mostly concerned with such questions as finding k given l and the figurate number f_k^l, and is of little interest. Next comes *De numerorum continuorum productis* concerning numbers of the form $l(l+1)(l+2)\ldots(l+k-1)$, about which Pascal manages to give several propositions arising directly from the fact that they are the numerators of figurate numbers defined by equation (1.18). Perhaps the most interesting observation is that, precisely because it gives rise to a figurate number, the product of k consecutive integers starting with l (as above) is necessarily divisible by $k!$, coupled with the remark that one could prove such results without using a knowledge of the Arithmetical Triangle, but the proofs would be longer and more tedious. Perhaps he had in mind proof by induction [21]. The next Treatise, *Numericarum potestatum generalis resolutio*, is essentially a continuation, of no particular interest to us, but then comes one we have already noticed: *Combinationes*, the first part of which is a Latin version of Part II Section 2 of *A Treatise on the Arithmetical Triangle*, described above, whilst the new part merely explores the properties of the Triangle in the language of combinations, and contains no surprises. Just how unwise it was of Montmort to conclude that Pascal's work was original "since he does not cite any other person" may be seen from Problem I of this Treatise: "Given any number, find the sum of all the combinations which can be made from it (without using the Arithmetical Triangle)". The solution is "In the 'doubling' progression starting with unity, take the term which is in order one greater than the number, and subtract unity from it". In other words, if n be the number, $2^n - 1$ is the solution, a very well-known result as we saw in Chapter 2.

Problem II of this Treatise uses the result given in Part II Section

2 of *A Treatise on the Arithmetical Triangle*

$$^nC_r = f_r^{n-r+1} \qquad\qquad (4.3\ bis)$$

and couples it with

$$f_k^l = \frac{(k+1)(k+2)(k+3)\ldots(k+l-1)}{(l-1)(l-2)(l-3)\ldots 1}, \qquad (6.20\ bis)$$

the chief result of Part I, to give, on putting $k = r$ and $l = nr + 1$ as usual,

$$^nC_r = \frac{(r+1)(r+2)(r+3)\ldots n}{(n-r)(n-r-1)(n-r-2)\ldots 1}. \qquad (7.11)$$

This, of course, is the Hindu rule (3.1) in the version $^nC_{n-r}$, and Pascal ends this Treatise with the following comment.

Writing that he has decided to finish the Treatise with this result ("not without regret, for I have many further results"), he adds that his friend M. de Gaignières communicated to him the rule

$$^nC_r = \frac{n(n-1)(n-2)\ldots(n-r+1)}{1.2.3\ldots r}, \qquad (3.1\ bis)$$

to which he had been led, with a request that Pascal find the demonstration of it. "I pondered over the problem, but, alarmed at its difficulty, I thought it proper to leave the demonstration to its author; however, thanks to the Arithmetical Triangle, an easy way was opened up". Pascal then observes, of course, that by the symmetry of the Triangle (1.18) could have been used with (4.3) to obtain (3.1), rather than (6.20) being used, which led to (7.11). In other words, Pascal recognizes the equivalence of the two forms given above, or that $^nC_r = ^nC_{n-r}$.

De Gaignières had certainly read Mersenne's combinatorial writings for he referred to them in a letter to Mersenne dated 17 February 1637 [22], and the impression that Pascal gives that de Gaignières invented the rule for himself must have been based on a misapprehension; besides, it is given by Hérigone (1634) [23]. It is interesting that Pascal failed to find a direct combinatorial demonstration of (3.1), in spite of having originated the combinatorial explanation,of the addition rule (7.1). The obvious one is via the combinatorial derivation of $^nC_r = ((n - r + 1)/r) \cdot {}^nC_{r-1}$ (4.4 bis) which we gave in Chapter 3 as the extension to Boethius's argument. Another possible demonstration is that provided by ben Gerson in 1321 (see our Chapter 4), but this uses the "n!" rule, which Pascal seems studiously to have avoided, in spite of its occurrence in the works of Mersenne and Hérigone [24]. It is almost as though he felt he had a mission to develop the theory of combinations without using it; he certainly never gave it.

The final two Treatises, *Potestatum numericarum summa* and *De numeris multiplicibus*, are thought to be older than the others [25]. The second concerns the divisibility of numbers as judged by the divisibility of the sums of their digits, and does not concern us, but the first describes a different recurrence formula for the sums of the powers of the integers from the one used by Fermat which we gave in Chapter 1. It is somewhat simpler, in that the coefficient of each sum is obtained explicitly, and it was destined to become the standard method given in algebra textbooks; it relies on the binomial expansion, and appears to be Pascal's own.

The result is

$$(n + 1)^r - (n + 1)$$

$$= r \sum n^{r-1} + \binom{r}{2} \sum n^{r-2} + \binom{r}{3} \sum n^{r-3} + \ldots + r \sum n, \quad (7.12)$$

where $\sum n^r$ is written for $\sum_{i=1}^{n} i^r$.

Applied recursively, the sums of the successive powers may be found. Pascal actually treated the more general case of the powers of the terms in any arithmetical progression, not only those starting with 1, but (7.12) suffices for our purposes. He proved the result for a particular case, and then observed that the method could be used generally. In essence, he argued as follows. Consider, as a particular example, the identity

$$(x + 1)^4 - x^4 = 4x^3 + 6x^2 + 4x + 1 \quad (7.13)$$

and put $x = 1, 2, 3, \ldots, n$ in it in turn, summing the results:

$$\left.\begin{array}{l} 2^4 - 1^4 = 4 \cdot 1^3 \quad + 6 \cdot 1^2 \quad + 4 \cdot 1 \quad + 1 \\[4pt] 3^4 - 2^4 = 4 \cdot 2^3 \quad + 6 \cdot 2^2 \quad + 4.2 \quad\;\; + 1 \\[4pt] 4^4 - 3^4 = 4 \cdot 3^3 \quad + 6 \cdot 3^2 \quad + 4 \cdot 3 \quad + 1 \\[6pt] \qquad\qquad\qquad \cdots \qquad\qquad\qquad \cdots \\[6pt] (n + 1)^4 - n^4 = 4 \cdot n^3 \quad + 6 \cdot n^2 \quad + 4 \cdot n \quad + 1 \end{array}\right\} \tag{7.14}$$

$$(n + 1)^4 - 1 \; = 4 \sum n^3 + 6 \sum n^2 + 4 \sum n + n.$$

The general result (7.12) is now easily seen by analogy.

Pascal, like Fermat, was interested in summing the powers of the integers in connection with quadrature, and ended this Treatise by observing that in the limit (using an arithmetical progression with a large number of terms and very small differences) one obtains, in modern notation,

$$\int_0^a x^k \, dx = \frac{a^{k+1}}{k + 1}, \tag{7.15}$$

which had already been given in a different form by Roberval [26] as well as by Fermat.

Although Pascal made no attempt to derive a general solution for n^{r-1} as a polynomial in n of degree r, which (7.12) obviously makes possible, it will pay us to do so in preparation for James Bernoulli's result on the sums of the powers (Chapter 10); with the advantage of hindsight afforded by matrix notation it is an easy matter [27].

Replacing n by $(n - 1)$ in (7.12) and using the equation successively for $r = 1, 2, 3, \ldots$ we find

$$\left.\begin{array}{l} n = n \\[4pt] n^2 = n + 2 \sum (n - 1) \\[4pt] n^3 = n + 3 \sum (n - 1) + 3 \sum (n - 1)^2 \\[4pt] n^4 = n + 4 \sum (n - 1) + 6 \sum (n - 1)^2 + 4 \sum (n - 1)^3 \end{array}\right\} \tag{7.16}$$

which may be written

$$
\begin{pmatrix} n \\ n^2 \\ n^3 \\ n^4 \\ \cdot \end{pmatrix} = \begin{pmatrix} 1 & 0 & 0 & 0 & \cdot \\ 1 & 2 & 0 & 0 & \cdot \\ 1 & 3 & 3 & 0 & \cdot \\ 1 & 4 & 6 & 4 & \cdot \\ \cdot & \cdot & \cdot & \cdot & \cdot \end{pmatrix} \begin{pmatrix} n \\ \sum (n-1) \\ \sum (n-1)^2 \\ \sum (n-1)^3 \\ \cdot \end{pmatrix}. \quad (7.17)
$$

Since the matrix of coefficients is triangular it may be readily inverted to give

$$
\begin{pmatrix} n \\ \sum (n-1) \\ \sum (n-1)^2 \\ \sum (n-1)^3 \\ \cdot \end{pmatrix} = \begin{pmatrix} 1 & 0 & 0 & 0 & \cdot \\ -1/2 & 1/2 & 0 & 0 & \cdot \\ 1/6 & -1/2 & 1/3 & 0 & \cdot \\ 0 & 1/4 & -1/2 & 1/4 & \cdot \\ \cdot & \cdot & \cdot & \cdot & \cdot \end{pmatrix} \begin{pmatrix} n \\ n^2 \\ n^3 \\ n^4 \\ \cdot \end{pmatrix}.
$$

$$(7.18)$$

To obtain the sums to n terms instead of $(n-1)$ we use $\sum n^r = \sum (n-1)^r + n^r$, which has the effect of adding 1's to the numbers immediately below the main diagonal of the matrix, turning each $-\frac{1}{2}$ into $+\frac{1}{2}$. The final result is then

$$
\begin{pmatrix} \sum 1 \\ \sum n \\ \sum n^2 \\ \sum n^3 \\ \cdot \end{pmatrix} = \begin{pmatrix} 1 & 0 & 0 & 0 & \cdot \\ 1/2 & 1/2 & 0 & 0 & \cdot \\ 1/6 & 1/2 & 1/3 & 0 & \cdot \\ 0 & 1/4 & 1/2 & 1/4 & \cdot \\ \cdot & \cdot & \cdot & \cdot & \cdot \end{pmatrix} \begin{pmatrix} n \\ n^2 \\ n^3 \\ n^4 \\ \cdot \end{pmatrix}. \quad (7.19)
$$

De Moivre was to call the numbers in the first column of the matrix the "Bernoulli numbers"; we shall continue the story of them in Chapter 10. How delighted Pascal would have been to learn that the complete solution to the problem of the sums of the powers of the integers could be obtained from his equation by the matrix inversion of his Arithmetical Triangle (omitting the final 1 in each row)! [28].

Notes to Chapter 7

[1] Leibniz gave (7.1) in 1666, a year after the publication of Pascal's *Traité*, and the result is often attributed to him (see Chapter 8).

[2] Hérigone (1634), *Arithmétique practique*, p. 122.

[3] Todhunter (1865).

[4] Montmort (1713), p. 4.

[5] Montmort (1713), p. xxxiii.

[6] See Chapter 6, note 6.

[7] Fermat to Pascal, 25 September 1654: "The consequence, as you so well remarked, of this fiction of lengthening the match to a particular number of games is that it serves only to simplify the rules and (in my opinion) to make all the chances equal or, to state it more intelligibly, to reduce all the fractions to the same denominator". For editions and translations of the correspondence, see Chapter 6, note 7.

[8] This description is based on Jacquard (1974), pp. 5–9.

[9] In Pascal's first surviving letter to Fermat, dated 29 July 1654, he wrote "Your method is very sound and is the one which first came to my mind in this research; but because the labour of the combinations is excessive, I have found a short cut and indeed another method which is much quicker and neater, . . ."; again, in the letter of 24 August 1654 Pascal wrote "when the first man needs *two* games and the second needs *three*, then to find the fair division of stakes, you say that one must know in how many games the play will be absolutely decided. It is easy to calculate that this will be in *four* games, from which you conclude that it is necessary to see in how many ways four games can be arranged between two players, and one must see how many combinations would make the first win and how many the second and to share out the stakes in proportion. I would have found it difficult to understand this if I had not known it myself already; in fact you had explained it with this idea in mind". It thus appears that although by the end of the correspondence Fermat identified with the method of *embedding* and subsequent *enumeration of the combinations*, and Pascal with his alternative (which we shortly describe in our account), originally they both used "Fermat's method". (Merrington's translations; see Chapter 6, note 7.)

[10] Pascal (1654b), Vol. II, p. 1147 footnote.

[11] Mahoney (1973), pp. 394–5. Mahoney assumes that Fermat knew that the numbers of games with 0, 1, 2, 3, . . . wins for A were given by the combinatorial numbers, but I see no evidence for this. Throughout the correspondence, it seems to me, counting combinations means simply enumerating all the 2^{a+b-1} games in full. Maistrov (1974), pp. 47–8, agrees.

[12] Montmort (1713), pp. 243–8, describes this solution as that "which the method of Fermat furnishes", but again (see the preceding note) there is no evidence that Fermat went beyond enumerating all the games.

[13] As we saw in Chapter 4, Levi ben Gerson used explicit mathematical induction (in 1321), and Maurolico (1575) (e.g. Book I, p. 7, Proposition 15) came very close to doing so. Pascal probably obtained the idea from the latter, since in 1659 he wrote of the proof of a proposition "*Cela est aisé par Maurolic*" (Pascal, 1654a, p. 137). Bussey (1917) gives an account of the contributions of both Maurolico and Pascal, but the priorities have been much debated, most recently by Hara (1962) and Rabinovitch (1970b).

[14] Sheynin (1977), p. 239.

[15] David (1962), pp. 70–97.
[16] Pascal (1654a), pp. 101–3; (1654b), Vol. II, pp. 1031–5. Mesnard (Pascal, 1654b, Vol. II, pp. 1019–31) argues convincingly that the *Académie Parisienne* is none other than Mersenne's Academy, but rather less convincingly that the letter was written in the early spring of 1654: for it refers to the fact that the problem *faire les partis des jeux* for two players has been solved, or at least that the solution is in sight, and we know that it was not solved until the autumn. July 1654 is a possibility. I am grateful to Professor Mesnard for discussing these questions with me.
[17] *A Dictionary of Statistical Terms* (Kendall and Buckland, 1971) says: "An unnecessary alternative name for the Negative Binomial Distribution, presumably because some untraced individual thought that Pascal discovered it". Montmort (1713), p. 245, gave an alternative solution to the Problem of Points which, though equivalent to (7.5), proceeds by direct calculation of the probabilities of the possible games rather than of the imaginary games of fixed length in which the real games may be embedded. Montmort thus introduced the negative binomial distribution (for he treated the general case, not just the case with a probability of one-half); Todhunter (1865), p. 98, believes this solution original to Montmort, after whom the negative binomial distribution might therefore be called.
[18] Hérigone (1634), *Algèbre*, pp. 16–19.
[19] This and the following quotations are from the letter to the *Académie Parisienne* (see note 16 above).
[20] See the discussion of Corollary 12 in Chapter 6.
[21] For which see Hardy and Wright (1954), p. 64.
[22] Mersenne (1648), pp. 190–9. (The reference is to a volume of Mersenne's correspondence; he died in 1648.)
[23] Hérigone (1634), *Arithmétique practique*, pp. 119–21.
[24] For Mersenne, see Chapter 4; Hérigone (1634), *Arithmétique practique*, pp. 123–4.
[25] According to Mesnard (Pascal, 1654b, Vol. II, p. 1170). Mesnard's argument relies heavily on his date for the letter to the *Académie Parisienne*, which I wish to place later for independent reasons (see note 16 above). Of course, the printing may be substantially later than the writing.
[26] Smith (1925), p. 688. On Pascal's formula in general, see Boyer (1943).
[27] Edwards (1982a).
[28] The computation of the matrix of (7.19), say $\{b_{ij}\}$, is easy: $b_{11} = 1$, b_{ij} $(j > i) = 0$, and all other values can be found from $b_{i+1, j+1} = (i/(j + 1))b_{ij}$ and the fact that each row sums to unity.

8

The Arithmetical Triangle in Analysis

In the third quarter of the seventeenth century the Arithmetical Triangle played a fundamental role in the development of analysis. Through the work of John Wallis it led Isaac Newton to the discovery of the binomial theorem for fractional and negative indices, and it was central to Leibniz's discovery of the calculus. Moreover it cropped up, in a more passive role, in the theory of finite differences as developed by Newton and Gregory (see Chapter 1).

Wallis [1], originally a Cambridge man but from 1649 to 1703 Savilian Professor of Geometry at Oxford, published his *Arithmetica infinitorum* [2] in 1655. Dedicated to William Oughtred [3], from whose *Clavis Mathematicae* he learnt of the Arithmetical Triangle, it culminates in the successful quadrature of the circle, Wallis finding that the area of a quadrant of the circle of unit radius (which is $\pi/4$) is given by the infinite product

$$\frac{2.4.4.6.6.8.8\ldots}{3.3.5.5.7.7.9\ldots}. \qquad (8.1)$$

This famous result was achieved by a speculative, but ultimately successful, interpolation of intermediate values in the Arithmetical Triangle, as follows.

Wallis's goal is the determination of

$$\frac{\pi}{4} = \int_0^1 (1 - x^2)^{1/2}\,dx, \qquad (8.2)$$

and he starts from the result

$$\int_0^a x^k \, dx = \frac{a^{k+1}}{k+1} \qquad (7.15 \; bis)$$

which Roberval, Fermat and Pascal had demonstrated to be true for positive integral values of k. His first assumption is that equation (7.15 bis) is true for all non-negative k, and in his Proposition 59 [4] he tabulates

$$\int_0^1 x^k \, dx = \frac{1}{k+1} \qquad (8.3)$$

for $k = p/q$, p and q being positive integers. This enables him to evaluate

$$\int_0^1 (1 - x^{1/p})^q \, dx \qquad (8.4)$$

for positive integral p, q by expansion and the use of (8.3). Thus, for example [5],

$$\int_0^1 (1 - x^{1/3})^2 \, dx = \int_0^1 (1 - 2x^{1/3} + x^{2/3}) \, dx$$

$$= \frac{1}{1} - \frac{2}{1 + \frac{1}{3}} + \frac{1}{1 + \frac{2}{3}} = \frac{1}{10}.$$

Wallis gives the corresponding tabulation in Proposition 131 [6] and observes, in a *scholium* [7], that he has arrived at a symmetrical table of inverse figurate numbers (such as Fig. 25).

If this is a tabulation of (8.4), argues Wallis, then the table of figurate numbers itself (which he gives in Proposition 132 [8], in Pascal's form complete with initial row and column of 1's, and with a mention of the addition rule of construction) must be a tabulation of

$$g(p, q) = 1 \bigg/ \int_0^1 (1 - x^{1/p})^q \, dx \qquad (8.5)$$

$$q$$

	1	2	3	4	5	6	
1	$\dfrac{1}{2}$	$\dfrac{1}{3}$	$\dfrac{1}{4}$	$\dfrac{1}{5}$	$\dfrac{1}{6}$	$\dfrac{1}{7}$	$\cdots$
2	$\dfrac{1}{3}$	$\dfrac{1}{6}$	$\dfrac{1}{10}$	$\dfrac{1}{15}$	$\dfrac{1}{21}$	$\dfrac{1}{28}$	$\cdots$
p 3	$\dfrac{1}{4}$	$\dfrac{1}{10}$	$\dfrac{1}{20}$	$\dfrac{1}{35}$	$\dfrac{1}{56}$	$\dfrac{1}{84}$	$\cdots$
4	$\dfrac{1}{5}$	$\dfrac{1}{15}$	$\dfrac{1}{35}$	$\dfrac{1}{70}$	$\dfrac{1}{126}$	$\dfrac{1}{210}$	$\cdots$
5	$\dfrac{1}{6}$	$\dfrac{1}{21}$	$\dfrac{1}{56}$	$\dfrac{1}{126}$	$\dfrac{1}{252}$	$\dfrac{1}{462}$	$\cdots$
6	$\dfrac{1}{7}$	$\dfrac{1}{28}$	$\dfrac{1}{84}$	$\dfrac{1}{210}$	$\dfrac{1}{462}$	$\dfrac{1}{924}$	$\cdots$
	$\vdots$	$\vdots$	$\vdots$	$\vdots$	$\vdots$	$\vdots$	

Fig. 25 The essence of Wallis's tabulation in his Proposition 131, which he recognized as a table of inverse figurate numbers

for $p, q, = 0, 1, 2, \ldots$. Thus Wallis assumes that

$$g(p, q) = f_p^{q+1}, \quad p, q = 0, 1, 2, \ldots, \qquad (8.6)$$

and this is indeed the case (for $p = 0$ it is true only in a limiting sense, and henceforth we interpret $g(0, q)$ accordingly).

Now in order to solve (8.2) Wallis needs to determine $g(\tfrac{1}{2}, \tfrac{1}{2})$, which he denotes by a small square symbol, and which we represent by w. Thus he has embedded his problem in the interesting problem (for us) of finding intermediate values in Pascal's Triangle according to some intuitively convincing rule.

The first stage is easy. In Propositions 171 to 182 [9] he establishes the fundamental representation of the figurate numbers (1.18) in the form

$$f^l_k = 1, \quad k = 0$$

$$f^l_k = \frac{l(l + 1)(l + 2) \ldots (l + k - 1)}{1 \cdot 2 \cdot 3 \ldots k}, \quad k = 1, 2, 3, \ldots$$

$$l = 1, 2, 3, \ldots. \qquad (8.7)$$

To conform with the present notation we must put $k = p$ and $l = q + 1$, whence

$$f^{q+1}_p = 1, \quad p = 0$$

$$f^{q+1}_p = \frac{(q + 1)(q + 2)(q + 3) \ldots (q + p)}{1 \cdot 2 \cdot 3 \ldots p}, \quad p = 1, 2, 3, \ldots$$

$$q = 0, 1, 2, \ldots. \qquad (8.8)$$

Wallis could have had these results from Briggs (himself the first holder of Wallis's Oxford professorship; see Chapter 1), but he gives the appearance of having worked them out for himself [10] by using the addition rule for the figurate numbers (1.7) and the known expressions for the sums of the squares, cubes, etc. In other words, he has reversed Fermat's procedure.

Wallis then assumes that the result derived from (8.6) and (8.8), namely

$$g(p, q) = 1, \quad p = 0$$

$$g(p, q) = \frac{(q + 1)(q + 2)(q + 3) \ldots (q + p)}{1 \cdot 2 \cdot 3 \ldots p}, \quad p = 1, 2, 3, \ldots$$

$$q = 0, 1, 2, \ldots, \qquad (8.9)$$

also holds for $q = -\frac{1}{2}, \frac{1}{2}, 1\frac{1}{2}, 2\frac{1}{2}, \ldots$, and in Proposition 184 he gives a table [11] (Fig. 26) with these halfway values entered in, and, by symmetry, the columns similarly treated. He notes that the addition rule of construction holds for these values. However, in the first row of Fig. 26 Wallis's intuition has momentarily deserted him, for although it is true that (8.9) does hold for the half-values quoted, and

Polynomials (column headings):

$$1 \qquad 1 \qquad \frac{l^2+l}{2} \qquad \frac{l^3+3l^2+2l}{6} \qquad \frac{l^4+6l^3+11l^2+6l}{24}$$

Triang. triang.	$\frac{105}{384}$	1	$21\frac{77}{384}$	5	$9\frac{9}{384}$	15	$23\frac{177}{384}$	35	$50\frac{105}{384}$	70
		1		$4\frac{1}{2}$		$12\frac{3}{8}$		$26\frac{39}{48}$		$50\frac{105}{384}$
Pyramidales.	$\frac{15}{48}$	1	$2\frac{9}{48}$	4	$6\frac{27}{48}$	10	$14\frac{21}{48}$	20	$26\frac{39}{48}$	35
		1		$3\frac{1}{2}$		$7\frac{7}{8}$		$14\frac{21}{48}$		$23\frac{177}{384}$
Triangulares.	$\frac{3}{8}$	1	$1\frac{7}{8}$	3	$4\frac{3}{8}$	6	$7\frac{7}{8}$	10	$12\frac{3}{8}$	15
		1		$2\frac{1}{2}$		$4\frac{3}{8}$		$6\frac{27}{48}$		$9\frac{9}{384}$
Laterales.	$\frac{1}{2}$	1	$1\frac{1}{2}$	2	$2\frac{1}{2}$	3	$3\frac{1}{2}$	4	$4\frac{1}{2}$	5
		1	$\square$	$1\frac{1}{2}$		$1\frac{7}{8}$		$2\frac{9}{48}$		$23\frac{177}{384}$
Monadici.	1	1	1	1	1	1	1	1	1	1
	∞	1		$\frac{1}{2}$		$\frac{3}{8}$		$\frac{15}{48}$		$\frac{105}{384}$
		Monadici.	Laterales.	Triangulares.	Pyramidales.					Trianguli-triang.

Et sic deinceps

Numeri

Fig. 26 The first stage in Wallis's interpolation of Pascal's Triangle. Note the "Stirling numbers of the first kind" in the polynomials

that

$$g(s, t) = g(t, s), \qquad s, t \geqslant 0, \tag{8.10}$$

it is not true that

$$g(-\tfrac{1}{2}, t) = g(t, -\tfrac{1}{2}),$$

since the former is not defined by (8.5). The first row of Fig. 26 therefore does not correspond to the function g, but happily this is of no consequence to the rest of Wallis's argument. (To keep the matter straight we now simply *define* $g(s, t) = g(t, s)$ for $-1 \leqslant s < 0, t \geqslant 0$.)

By this means Wallis has succeeded in filling some, but not all, of the halfway values. The squares intermediate to both rows *and* columns of Pascal's Triangle are still empty, including the one at $(\tfrac{1}{2}, \tfrac{1}{2})$ which is the ultimate prey; Wallis fills it with a small square symbol (note that at $(-\tfrac{1}{2}, -\tfrac{1}{2})$ he has inserted "∞", the first use of this symbol [12]). Wallis now sets out to trap his prey.

In Proposition 182 he has already observed that to obtain the series of figurate numbers for $(p + 1)$ from that for p it is only necessary to write (8.8) as

$$f_{p+1}^{q+1} = f_p^{q+1} \cdot \frac{q + p + 1}{p + 1}, \tag{8.11}$$

which is of course another of Briggs's results (1.19), and he now decides (Proposition 187 [13]) that the similar result for the function g,

$$g(p + 1, q) = g(p, q) \cdot \frac{q + p + 1}{p + 1}, \tag{8.12}$$

applies not only for the integral values of p but also to the half values $p = -\tfrac{1}{2}, \tfrac{1}{2}, 1\tfrac{1}{2}, 2\tfrac{1}{2}, \ldots$. Once again he is right (with the assumption we made above concerning $g(-\tfrac{1}{2}, q)$).

Applying (8.12), Wallis is able (Proposition 188) to compute the numbers needed to fill out Fig. 26, each as a multiple of the number $g(-\tfrac{1}{2}, q)$ at the head of its column. Since by (8.12)

$$w = g(\tfrac{1}{2}, \tfrac{1}{2}) = g(-\tfrac{1}{2}, \tfrac{1}{2}) \cdot 2, \tag{8.13}$$

$g(-\frac{1}{2}, \frac{1}{2}) = w/2$, whence $g(\frac{1}{2}, -\frac{1}{2}) = w/2$ and the rest of the first column follows. This also gives the rest of the first row, and hence the rest of the table, all the new values being multiples of w. The final table is given in Proposition 189 [14] and our Fig. 27.

The way is now clear for the determination of w: by concentrating on the row for $p = \frac{1}{2}$ (which he gives in the form of Fig. 28 in his Proposition 191 [15]), Wallis assumes a certain smoothness in the successive values, not only that they increase monotonically [16], but that the ratio of each to its predecessor decreases monotonically. Thus for three consecutive terms a, b and c,

$$\frac{c}{b} < \frac{b}{a} \quad \text{or} \quad ac < b^2. \tag{8.14}$$

Applying this to the last triplet and to the penultimate triplet in Fig. 28, for example, we have, after a little rearrangement,

$$\left.
\begin{aligned}
\frac{3.3.5.5.7.7}{2.4.4.6.6.8}\sqrt{\frac{9}{8}} &< w \\[2ex]
\text{and} \\[2ex]
\frac{3.3.5.5.7.7}{2.4.4.6.6.8}\sqrt{\frac{8}{7}} &> w
\end{aligned}
\right\} \tag{8.15}$$

These products may be continued indefinitely, and so Wallis arrives at

$$w = \frac{4}{\pi} = \frac{3.3.5.5.7.7\ldots}{2.4.4.6.6.8\ldots}. \tag{8.16}$$

Wallis has thus solved his problem by assuming that the function $g(p, q)$ behaves for non-integral values of its arguments in the same way as it does for integral values. By noting the correspondence between g and the figurate numbers, by noting that the formulae for the figurate numbers can be employed for non-integral arguments, by assuming symmetry, and, finally, by assuming a certain smoothness, he has brilliantly succeeded in trapping $w = 4/\pi$. All his speculations about g turn out to be true for non-negative arguments.

∞	1	w	$\tfrac{1}{2}$	$\tfrac{1}{3}w$	$\tfrac{3}{8}$	$\tfrac{4}{15}w$	$\tfrac{15}{48}$	$\tfrac{8}{35}w$	$\tfrac{105}{384}$	A
1	1	1	1	1	1	1	1	1	1	1
$\tfrac{1}{2}w$	1	$\tfrac{3}{2}$	2	$\tfrac{4}{3}$	$\tfrac{15}{8}$	$\tfrac{8}{5}$	$\tfrac{105}{48}$	$\tfrac{64}{35}$	$\tfrac{945}{384}$	$A \times \dfrac{2l-1}{1}$
$\tfrac{1}{2}$	1	$\tfrac{3}{2}$	2	$\tfrac{5}{2}$	3	$\tfrac{7}{2}$	4	$\tfrac{9}{2}$	5	$l = \dfrac{2l \pm 0}{2}$
$\tfrac{1}{3}w$	1	$\tfrac{4}{3}$	$\tfrac{5}{2}$	$\tfrac{8}{3}$	$\tfrac{35}{8}$	$\tfrac{64}{15}$	$\tfrac{315}{48}$	$\tfrac{128}{21}$	$\tfrac{3465}{384}$	$A \times \dfrac{4l^2-1}{3}$
$\tfrac{3}{8}$	1	$\tfrac{15}{8}$	3	$\tfrac{35}{8}$	6	$\tfrac{63}{8}$	10	$\tfrac{99}{8}$	15	$\dfrac{l^2+1}{2} = \dfrac{4l^2+4l}{8}$
$\tfrac{4}{15}w$	1	$\tfrac{8}{5}$	$\tfrac{7}{2}$	$\tfrac{64}{15}$	$\tfrac{63}{8}$	$\tfrac{128}{15}$	$\tfrac{693}{48}$	$\tfrac{512}{35}$	$\tfrac{9009}{384}$	$A \times \dfrac{8l^3+12l^2-2l-3}{15}$
$\tfrac{15}{48}$	1	$\tfrac{105}{48}$	4	$\tfrac{315}{48}$	10	$\tfrac{693}{48}$	20	$\tfrac{1287}{48}$	35	$\dfrac{l^3+3l^2+2l}{6} = \dfrac{8l^3+24l^2+16l}{48}$
$\tfrac{8}{35}w$	1	$\tfrac{64}{35}$	$\tfrac{9}{2}$	$\tfrac{128}{21}$	$\tfrac{99}{8}$	$\tfrac{512}{35}$	$\tfrac{1287}{48}$	$\tfrac{1024}{35}$	$\tfrac{19305}{384}$	$A \times \dfrac{16l^4+64l^3+56l^2-16l-15}{105}$
$\tfrac{105}{384}$	1	$\tfrac{945}{384}$	5	$\tfrac{3465}{384}$	15	$\tfrac{9009}{384}$	35	$\tfrac{19305}{384}$	70	$\dfrac{l^4+6l^3+11l^2+6l}{24} = \dfrac{16l^4+96l^3+176l^2+96l}{384}$

Fig. 27 The second stage in Wallis's interpolation of Pascal's Triangle. A refers generically to the number at the head of the column; Wallis's symbol has been replaced by w

q	$-\frac{1}{2}$	0	$\frac{1}{2}$	1	$1\frac{1}{2}$	2	$2\frac{1}{2}$	3	$3\frac{1}{2}$	4	...
$p=\frac{1}{2}$	$\frac{1}{2}w$	1	w	$\frac{3}{2}$	$\frac{4}{3}w$	$\frac{3.5}{2.4}$	$\frac{4.6}{3.5}w$	$\frac{3.5.7}{2.4.6}$	$\frac{4.6.8}{3.5.7}w$	$\frac{3.5.7.9}{2.4.6.8}$	...

Fig. 28 The completed row for $p=\frac{1}{2}$

In his study of the function he has shown how to compute it not only for any q when p is integral [17], and vice versa by symmetry, but also for both non-integral; for his method of determining g for half-integral values could be repeated for quarter-integral values, and so on indefinitely. It may fairly be said, therefore, that Wallis has invented the complete beta-function, as follows.

In (8.5) put $y = x^{1/p}$, $p > 0$, whence

$$\frac{1}{g(p,q)} = \int_0^1 (1-y)^q p y^{p-1}\,\mathrm{d}y = pB(p, q+1). \qquad (8.17)$$

In 1729 Euler was to introduce the gamma-function in the course of interpolating the series 1!, 2!, 3!, ..., [18] and connect it with Wallis's function, but the essential step of finding the smooth function which interpolates Pascal's Triangle was Wallis's, even though the discovery was, as we have seen, actually made the other way round, by studying the function and noticing that it included the figurate numbers as special values.

To understand how Wallis's work led Isaac Newton to discover the binomial theorem for non-integral index in 1664/65 we return to (8.2),

$$\frac{\pi}{4} = \int_0^1 (1-x^2)^{1/2}\,\mathrm{d}x, \qquad (8.2\ bis)$$

and let Newton tell the story himself in a letter written in 1676 [19].

"At the beginning of my mathematical studies, when I had met with the works of our celebrated Wallis, on considering the series by the intercalation of which he himself exhibits the area of the circle and the hyperbola, the fact that, in the series of curves whose common base or axis is x and the ordinates

$$(1-x)^{\frac{0}{2}}, \quad (1-x^2)^{\frac{1}{2}}, \quad (1-x^2)^{\frac{2}{2}}, \quad (1-x^2)^{\frac{3}{2}}, \quad (1-x^2)^{\frac{4}{2}}, \quad (1-x^2)^{\frac{5}{2}},$$

etc.,

if the areas of every other of them, namely

$$x, \quad x - \tfrac{1}{3}x^3, \quad x - \tfrac{2}{3}x^3 + \tfrac{1}{5}x^5, \quad x - \tfrac{3}{3}x^3 + \tfrac{3}{5}x^5 - \tfrac{1}{7}x^7, \quad \text{etc.,}$$

could be interpolated, we should have the areas of the intermediate ones, of which the first $(1-x^2)^{\frac{1}{2}}$ is the circle: in order to interpolate these series I noted that in all of them the first term was x and that the second terms $\frac{0}{3}x^3, \frac{1}{3}x^3, \frac{2}{3}x^3, \frac{3}{3}x^3$, etc., were in arithmetical progression, and hence that the first two terms of the series to be intercalated ought to be $x - \tfrac{1}{3}(\tfrac{1}{2}x^3)$, $x - \tfrac{1}{3}(\tfrac{3}{2}x^3)$, $x - \tfrac{1}{3}(\tfrac{5}{2}x^3)$, etc. To intercalate the rest I began to reflect that the denominators 1, 3, 5, 7, etc. were in arithmetical progression, so that the numerical coefficients of the numerators only were still in need of investigation. But in the alternately given areas these were the figures of powers of the number 11, namely of these, 11^0, 11^1, 11^2, 11^3, 11^4, that is, first 1; then 1, 1; thirdly, 1, 2, 1; fourthly 1, 3, 3, 1; fifthly 1, 4, 6, 4, 1, etc. And so I began to inquire how the remaining figures in these series could be derived from the first two given figures, and I found that on putting m for the second figure, the rest would be produced by continual multiplication of the terms of this series,

$$\frac{m-0}{1} \times \frac{m-1}{2} \times \frac{m-2}{3} \times \frac{m-3}{4} \times \frac{m-4}{5}, \text{ etc.}$$

For example, let $m = 4$, and $4 \times \tfrac{1}{2}(m-1)$, that is 6 will be the third term, and $6 \times \tfrac{1}{3}(m-2)$, that is 4 the fourth, and $4 \times \tfrac{1}{4}(m-3)$, that is 1 the fifth, and $1 \times \tfrac{1}{5}(m-4)$, that is 0 the sixth, at which term in this case the series stops. Accordingly, I applied this rule for interposing series among series, and since, for the circle, the second term was $\tfrac{1}{3}(\tfrac{1}{2}x^3)$, I put $m = \tfrac{1}{2}$, and the terms arising were

$$\frac{1}{2} \times \frac{\frac{1}{2} - 1}{2} \quad \text{or} \quad -\frac{1}{8}, \qquad -\frac{1}{8} \times \frac{\frac{1}{2} - 2}{3} \quad \text{or} \quad +\frac{1}{16},$$

$$\frac{1}{16} \times \frac{\frac{1}{2} - 3}{4} \quad \text{or} \quad -\frac{5}{128},$$

and so to infinity. Whence I came to understand that the area of the circular segment which I wanted was

$$x - \frac{\frac{1}{2}x^3}{3} - \frac{\frac{1}{8}x^5}{5} - \frac{\frac{1}{16}x^7}{7} - \frac{\frac{5}{128}x^9}{9} \quad \text{etc.}$$

And by the same reasoning the areas of the remaining curves, which were to be inserted, were likewise obtained: as also the area of the hyperbola and of the other alternate curves in this series $(1 + x^2)^{\frac{0}{2}}, (1 + x^2)^{\frac{1}{2}}, (1 + x^2)^{\frac{2}{2}}, (1 + x^2)^{\frac{3}{2}}$, etc. And the same theory serves to intercalate other series, and that through intervals of two or more terms when they are absent at the same time. This was my first entry upon these studies, and it had certainly escaped my memory, had I not a few weeks ago cast my eye back on some notes.

But when I had learnt this, I immediately began to consider that the terms

$$(1 - x^2)^{\frac{0}{2}}, \quad (1 - x^2)^{\frac{2}{2}}, \quad (1 - x^2)^{\frac{4}{2}}, \quad (1 - x^2)^{\frac{6}{2}}, \quad \text{etc.,}$$

that is to say,

$$1, \quad 1 - x^2, \quad 1 - 2x^2 + x^4, \quad 1 - 3x^2 + 3x^4 - x^6, \quad \text{etc.,}$$

could be interpolated in the same way as the areas generated by them: and that nothing else was required for this purpose but to omit the denominators 1, 3, 5, 7, etc., which are in the terms expressing the areas; this means that the coefficients of the terms of the quantity to be intercalated $(1 - x^2)^{\frac{1}{2}}$, or $(1 - x^2)^{\frac{3}{2}}$, or in general $(1 - x^2)^m$, arise by the continued multiplication of the terms of this series

$$m \times \frac{m-1}{2} \times \frac{m-2}{3} \times \frac{m-3}{4}, \text{ etc.,}$$

so that (for example)

$$(1-x^2)^{\frac{1}{2}} \text{ was the value of } 1 - \tfrac{1}{2}x^2 - \tfrac{1}{8}x^4 - \tfrac{1}{16}x^6 \text{ etc.,}$$

$$(1-x^2)^{\frac{3}{2}} \text{ of } 1 - \tfrac{3}{2}x^2 + \tfrac{3}{8}x^4 + \tfrac{1}{16}x^6, \text{ etc.,}$$

and

$$(1-x^2)^{\frac{1}{3}} \text{ of } 1 - \tfrac{1}{3}x^2 - \tfrac{1}{9}x^4 - \tfrac{5}{81}x^6, \text{ etc.}$$

So then the general reduction of radicals into infinite series by that rule, which I laid down at the beginning of my earlier letter became known to me, and that before I was acquainted with the extraction of roots. But once this was known, that other could not long remain hidden from me. For in order to test these processes, I multiplied

$$1 - \tfrac{1}{2}x^2 - \tfrac{1}{8}x^4 - \tfrac{1}{16}x^6, \quad \text{etc.}$$

into itself; and it became $1 - x^2$, the remaining terms vanishing by the continuation of the series to infinity. And even so $1 - \tfrac{1}{3}x^2 - \tfrac{1}{9}x^4 - \tfrac{5}{81}x^6$, etc. multiplied twice into itself also produced $1 - x^2$. And as this was not only sure proof of these conclusions so too it guided me to try whether, conversely, these series, which it thus affirmed to be roots of the quantity $1 - x^2$, might not be extracted out of it in an arithmetical manner. And the matter turned out well. This was the form of the working in square roots.

$$1 - x^2(1 - \tfrac{1}{2}x^2 - \tfrac{1}{8}x^4 - \tfrac{1}{16}x^6, \quad \text{etc.}$$

$$\frac{1}{0-x^2}$$

$$\frac{-x^2 + \tfrac{1}{4}x^4}{-\tfrac{1}{4}x^4}$$

$$\frac{-\tfrac{1}{4}x^4 + \tfrac{1}{8}x^6 + \tfrac{1}{64}x^8}{0 \quad -\tfrac{1}{8}x^6 - \tfrac{1}{64}x^8} \text{,,}$$

Newton had not only "met with the works of our celebrated Wallis"; he had steeped himself in them, and the notes over which he had cast his eye "a few weeks ago" are headed "Annotations out of Dr Wallis his Arithmetica infinitorum" [20].

First we observe that Newton is treating a significantly more general problem than Wallis, for he is attempting to find the *indefinite* integral in each case, for example

$$\int_0^x (1 - x^2)^{\frac{1}{2}} \, \mathrm{d}x \qquad (8.18)$$

(compare 8.2). He knows the answers for the cases with integral exponent (because of 7.15) and intends to find the series for half-integral exponents by an interpolation of series, for which purpose he sets out a table [21] (Fig. 29, upper part) which summarizes his knowledge of the series for integral exponents. As he tells us in his letter, he has noticed that the numerators of the coefficients are the elements of the Arithmetical Triangle. (We may note that in Newton's case the Triangle manifests itself immediately, whereas Wallis, treating a definite integral with two arguments (8.5), encountered it in inverse form and consequently had to invert his function. The two occurrences are similar, but not the same.) It was Newton's good fortune to be considering a problem which turned up a pattern of coefficients with which his reading of Wallis (and Oughtred) had made him thoroughly familiar [22].

From Fig. 29 we see that for Newton's problem, unlike Wallis's, the natural format for the Arithmetical Triangle is not Pascal's but a variant in which the binomial coefficients for each exponent are arrayed in columns (cf. Montmort's form, Fig. 1). Each column corresponds to the series for a particular exponent, and each row to a particular power of x.

Newton now proceeded to do two things. The first (I suggest) was to interpolate the half values of the Triangle in true Wallisian style, and the second was to discover for himself the columnwise rule of formation he quotes in his letter, and to observe that it worked for fractional and negative arguments. In the letter he does not describe the first stage, and indeed there is no good reason why he should: it is entirely natural to go straight to a description of the substantive result.

1st. $+x \times 1$.   1.   1.   1.   1.   1. 1. 1. 1. 1. 1.

2^{d}. $-\dfrac{x^3}{3} \times 0$. $0+1=1$. $1+1=2$. $2+1=3$. $3+1=4$. $4+1=5$. 6. 7. 8. 9. 10.

3^{d}. $+\dfrac{x^5}{5} \times 0$. $0+0=0$. $0+1=1$. $1+2=3$. $3+3=6$. $6+4=10$ 15. 21. 28. 36. 45.

4th, $-\dfrac{x^7}{7} \times 0$. $0+0=0$. $0+0=0$. $0+1=1$. $1+3=4$. $4+6=10$ 20. 35. 56. 84. 120.

5. $+\dfrac{x^9}{9} \times 0$. $0+0=0$. $0+0=0$. $0+0=0$. $0+1=1$. $1+4=5$. 15. 35. 70. 126. 210.

6. $-\dfrac{x^{11}}{11} \times 0$. $0+0=0$. $0+0=0$. $0+0=0$. $0+0=0$. $0+1=1$. 6. 21. 56. 126. 252.

1st. *. 2^{d}. *. 3^{d}. *. 4th. *. 5^{t}. *. 6^{t}. 1. 7. 28. 84. 210.

*.

7th 1. 8. 36. 120.

*.

8th 1. 9. 45.

*.

Now if the meane termes in these progressions
can bee calculated y^e first of y^m gives y^e
area *aeqp*. Which is thus done.

9th 1. 10.

*.

10th 1.

*.

11^{tl}

1st $+x$ $\times 1$. 1. 1. 1. 1. 1. 1. 1. 1. 1. 1.

2^{d}. $-\dfrac{x^3}{3} \times 0$. $\dfrac{1}{2}$. 1. $\dfrac{3}{2}$. 2. $\dfrac{5}{2}$. 3. $\dfrac{7}{2}$. 4. $\dfrac{9}{2}$. 5. $\dfrac{11}{2}$. 6.

3^{d}. $+\dfrac{1}{5}x^5 \times 0$. $-\dfrac{1}{8}$. 0. $\dfrac{3}{8}$. 1. $\dfrac{15}{8}$. 3. $\dfrac{35}{8}$. 6. $\dfrac{63}{8}$.10. $\dfrac{99}{8}$. 15.

4. $-\dfrac{1}{7}x^7 \times 0$. $+\dfrac{1}{16}$. 0. $-\dfrac{1}{16}$. 0. $\dfrac{5}{16}$. 1. $\dfrac{35}{16}$. 4. $\dfrac{105}{16}$. 10. $\dfrac{231}{16}$. 20.

5. $+\dfrac{1}{9}x^9 \times 0$. $-\dfrac{5}{128}$. 0. $\dfrac{3}{128}$. 0. $\dfrac{-5}{128}$. 0. $\dfrac{35}{128}$. 1. $\dfrac{315}{128}$. 5. $\dfrac{1155}{128}$. 15.

6. $-\dfrac{1}{11}x^{11} \times 0$. $\dfrac{7}{256}$. 0. $\dfrac{-3}{256}$. 0. $\dfrac{3}{256}$. 0. $\dfrac{-7}{256}$. 0. $\dfrac{63}{256}$. 1. $\dfrac{693}{256}$. 6.

7. $+\dfrac{1}{13}x^{13} \times 0$. $\dfrac{-21}{1024}$. 0. $\dfrac{7}{1024}$. 0. $\dfrac{-5}{1024}$. 0. $\dfrac{7}{1024}$. 0. $\dfrac{-21}{1024}$. 0. $\dfrac{231}{1024}$. 1. $\dfrac{3003}{1024}$.

Fig. 29 Pascal's Triangle interpolated by Newton during his discovery of the general binomial theorem. In the lower table the line (added by the present author) marks the limit of the intermediate values already calculated by Wallis

The first stage is straightforward, given the form of the upper table in Fig. 29 and the intention to make a Wallisian interpolation. For the half values in the upper right-hand part of the lower table have already been given by Wallis in several tables (Fig. 26 and 27), so that by straight transcription Newton could have filled in the table down to the line I have added. Then, by the addition rule already announced by Wallis to hold for the half values [23], Newton fills in the first five rows of the table leftwards [24]. (The zeros, of course, are straightforward.) To obtain the sixth and seventh rows Newton will (I believe) have used Wallis's Proposition 182 (equation 8.11 above) to obtain the last two values in each row,

$$\frac{693}{256} = \frac{315}{128} \times \frac{11/2}{5}$$

$$\text{and} \quad \frac{3003}{1024} = \frac{693}{256} \times \frac{13/2}{6},$$

and then continued leftwards using the addition rule. The advantage of this explanation is that it accounts for the "hanging" $\frac{3003}{1024}$ in the table and it leads directly to the general columnwise rule, as follows.

Wallis had given, for the series of figurate numbers (Fig. 26),

Monadici	1
Laterales	l
Triangulares	$\dfrac{l}{1} \cdot \dfrac{l+1}{2}$
Pyramidales	$\dfrac{l}{1} \cdot \dfrac{l+1}{2} \cdot \dfrac{l+2}{3}$

etc.,

and Newton has rearranged the series by shifting the *Triangulares* one step to the right, the *Pyramidales* two, and so on. So if in the *Laterales* we replace l by m (to conform to Newton's notation), in the *Triangulares* we replace l by $m - 1$, in the *Pyramidales* l by $m - 2$, and so on, we will have achieved the transformation from the Wallisian to the Newtonian form of the Arithmetical Triangle.

Running down the Newtonian column which starts 1, m, it continues

$$\frac{m}{1} \times \frac{m-1}{2}, \quad \frac{m}{1} \times \frac{m-1}{2} \times \frac{m-2}{3}, \quad \ldots,$$

just as Newton asserts. Moreover, there is no difficulty in applying these expressions to non-integral m, for Wallis had already used his forms of them with non-integral l. Newton has not only rediscovered the familiar multiplicative form (5.2) [25], but has applied it with non-integral argument. As we saw in Chapter 1, Harriot had made the identical rearrangement fifty years previously.

With this explained, we may return to Newton's letter for a description of how he obtained the binomial theorem itself, by realizing that the original functions "could be interpolated in the same way as the areas generated by them", and how he checked the emerging infinite series for $(1 - x^2)^{\frac{1}{2}}$, first by squaring it and then by extracting the square root directly "in an arithmetical manner". (The "earlier letter" to which he refers had quoted the binomial theorem [26].)

A while later Newton redrafted his work, extending Pascal's Triangle to the left (to correspond to negative exponents, obtaining another of Harriot's forms) and showing how to interpolate the rows of his form of Pascal's Triangle at one-third intervals. Wallis had already done this, and Newton ends the redraft by recalling Wallis's method, that any intermediate values can be computed from the expressions for the figurate numbers.

If this outline of the discovery of the general binomial theorem is approximately correct, we can see how the crucial element was Newton's desire to work with the indefinite integral, for, given the happy circumstance that the development was to reveal a familiar number pattern, once he had adopted the tabular form of working, the remaining steps—though there are a number of them—follow with a certain ease. The existence of non-zero interpolates amongst the zeros of Fig. 29 may momentarily have seemed awkward to Newton, but they will have surprised Wallis much more: in his first account of the general binomial theorem [27] in 1685 Wallis owned that he had earlier contemplated the possibility of interpolating a series between (for example)

$$(a + e)^2 = a^2 + 2ae + e^2 \quad \text{and} \quad (a + e)^3 = a^3 + 3a^2e + 3ae^2 + e^3$$

to represent $(a + e)^{2\frac{1}{2}}$, but since it would presumably contain more terms than three but fewer than four he "gave it over as a thing not feasible. That is, not a terminated Proportion". He then quotes the multiplicative formula Newton uses [28] "(what I had formerly sought after, but unsuccessfully)". Here Wallis seems to be admitting his ignorance of the multiplicative formula (5.2) even for integral powers, which is strange, for even if he had not read Briggs, given the formulae for the figurate numbers and the determination to find its equivalent for the binomial numbers, the transformation is simple by the standards of Wallis's other work.

James Gregory independently discovered the general binomial expansion in November 1670 [29]. Formal proofs of the theorem were lacking for many years, being ultimately furnished by Euler for rational exponents and Abel for the completely general case [30]. On its influence in the development of the calculus by Newton we need hardly comment; by leading the way to it, Pascal's Arithmetical Triangle played a vital role, a role which, as we see next, was paralleled in Leibniz's own discovery of the calculus.

Gottfried Wilhelm Leibniz was born in Leipzig in 1646. He delivered the *Disputatio arithmetica de complexionibus* in March 1666 as a new Master of Arts at the university there [31], and it was printed later that year in enlarged form as the *Dissertatio de arte combinatoria* [32]. Leibniz was not then aware of Pascal's *Traité*, which had appeared the previous year, and apparently drew his inspiration largely from a book of mathematical recreations by Schwenter [33]. From our point of view there are three noteworthy results in the *Dissertatio*. First, the direct combinatorial proof of the addition relation in the Arithmetical Triangle (7.1), which thus appeared under his name only a year after it had appeared under Pascal's; secondly, a table explaining how the ten combinations of five things taken three at a time ($^5C_3$) can be derived from the combinations of two, three, and four things taken two at a time ($^2C_2 + {}^3C_2 + {}^4C_2$); and thirdly the observation (without proof) that if n is prime it divides $\binom{n}{r}$ for all r between 1 and $n - 1$, a result observed for $n = 7$ by Rabbi ben Ezra in about 1140 [34]. The importance of this early work lies in the fact that when Leibniz arrived in Paris in 1672 his familiarity with the Arithmetical

Triangle enabled him to generalize his solution to a problem put to him by Huygens in a way which, by his own account, provided the insight into the inverse relationship between quadrature and finding the tangent which was crucial to his development of the calculus [35].

Before he met Huygens, Leibniz developed a theorem on the summation of consecutive terms of a series of differences. Consider the sequence

$$a_0, a_1, a_2, \ldots, a_n$$

and its first differences $d_i = a_i - a_{i-1}$, $i = 1, 2, 3, \ldots, n$. Then

$$d_1 + d_2 + d_3 + \ldots + d_n = (a_1 - a_0) + (a_2 - a_1) + \ldots + (a_n - a_{n-1})$$
$$= a_n - a_0, \tag{8.19}$$

or *the sum of the consecutive differences equals the difference between the first and last terms of the original sequence.* Thus, since the differences of the squares

$$0, 1, 4, 9, 16, \ldots$$

are the consecutive odd numbers

$$1, 3, 5, 7, 9, \ldots,$$

it follows that the sum of the first n odd numbers is n^2.

This theorem persuaded Leibniz that it ought to be possible to sum any regular series, even one with an infinite number of terms (provided the sum approaches a finite limit). On meeting Huygens, Leibniz told him of his result, whereupon Huygens suggested that he try to determine the sum of the infinite series of the reciprocals of the triangular numbers

$$\frac{1}{1} + \frac{1}{3} + \frac{1}{6} + \frac{1}{10} + \ldots \tag{8.20}$$

which had cropped up in a discussion on games of chance with Hudde in 1665 and which Huygens had found to be 2.

Leibniz noticed that the terms of this series are twice the differences of the harmonic series, thus:

$$\frac{1}{1} \quad \frac{1}{2} \quad \frac{1}{3} \quad \frac{1}{4} \quad \frac{1}{5} \quad \cdots$$

differences $\qquad \frac{1}{2} \quad \frac{1}{6} \quad \frac{1}{12} \quad \frac{1}{20}$ $\hfill$ (8.21)

twice
differences $\qquad \frac{1}{1} \quad \frac{1}{3} \quad \frac{1}{6} \quad \frac{1}{10} \quad \cdots$

whence, by his theorem, the sum required must be $2(1 - 0) = 2$. Since the nth triangular number is $\frac{1}{2}n(n + 1)$, Leibniz has effectively used

$$\frac{2}{n(n + 1)} = 2\left(\frac{1}{n} - \frac{1}{n + 1}\right), \qquad (8.22)$$

from which, on writing out a few terms, it is obvious that all but the first and last terms cancel, a trick which Leibniz could have noted in Pascal's method of summing the powers of the integers (7.14).

Huygens is reported to have been delighted at the method, which Leibniz went on to generalize by considering the inverse of other numbers in the Arithmetical Triangle. We give his table in Fig. 30 [36]. We need not comment on the first two columns; the sum of the third—the harmonic series itself—had been known to be infinite since the fourteenth-century demonstration by Nicole Oresme [37], whilst, unbeknown to Leibniz at this time, the sums of the fourth and the fifth columns had been published by Mengoli in 1650 [38]. These last two, and all the others, may be most immediately derived by considering the related triangle of Fig. 31. This is obtained by omitting the first two columns of Leibniz's Triangle and then dividing the kth remaining column by k, and possesses the property that each number is the difference between the number immediately above and the one above and to the right (starting with the harmonic series as the first row), whence each number is the sum of all the terms in the row below it starting with the term immediately below and proceeding to the right, by Leibniz's theorem. It was this inverse of the relationships which exist amongst the numbers of the Figurate Triangle (cf. equations 1.6 and 1.7) which impressed on

0	1	2	3	4	5	·	·

$\dfrac{0}{0}$	$\dfrac{1}{1}$	$\dfrac{1}{1}$	$\dfrac{1}{1}$	$\dfrac{1}{1}$	$\dfrac{1}{1}$	·	·
$\dfrac{0}{0}$	$\dfrac{1}{1}$	$\dfrac{1}{2}$	$\dfrac{1}{3}$	$\dfrac{1}{4}$	$\dfrac{1}{5}$	·	·
$\dfrac{0}{0}$	$\dfrac{1}{1}$	$\dfrac{1}{3}$	$\dfrac{1}{6}$	$\dfrac{1}{10}$	$\dfrac{1}{15}$	·	·
$\dfrac{0}{0}$	$\dfrac{1}{1}$	$\dfrac{1}{4}$	$\dfrac{1}{10}$	$\dfrac{1}{20}$	$\dfrac{1}{35}$	·	·
·	·	·	·	·	·	·	·
·	·	·	·	·	·	·	·

Sum $\quad \dfrac{0}{0} \quad \dfrac{1}{0} \quad \dfrac{1}{0} \quad \dfrac{2}{1} \quad \dfrac{3}{2} \quad \dfrac{4}{3}$

Fig. 30   Leibniz's Triangle of inverse figurate numbers

$$\dfrac{1}{1} \quad \dfrac{1}{2} \quad \dfrac{1}{3} \quad \dfrac{1}{4} \quad \dfrac{1}{5} \quad \dfrac{1}{6} \quad \cdot \quad \cdot$$

$$\dfrac{1}{2} \quad \dfrac{1}{6} \quad \dfrac{1}{12} \quad \dfrac{1}{20} \quad \dfrac{1}{30} \quad \cdot \quad \cdot$$

$$\dfrac{1}{3} \quad \dfrac{1}{12} \quad \dfrac{1}{30} \quad \dfrac{1}{60} \quad \cdot \quad \cdot$$

$$\dfrac{1}{4} \quad \dfrac{1}{20} \quad \dfrac{1}{60} \quad \cdot \quad \cdot$$

$$\cdot \quad \cdot \quad \cdot \quad \cdot$$

$$\cdot \quad \cdot \quad \cdot$$

Fig. 31   The Harmonic Triangle formed by successive differences of the harmonic
series in the first row

Leibniz's mind the fact that taking differences and forming sums are
reciprocal operations, a view so influential in his development of the
calculus.

Leibniz continued to work on combinatorial problems all his life
[39]. Sometime before August 1673 he wrote a note *De numeris
combinatoriis* [40], the most interesting observation in which is the
relation

$$f_k^l = 1 \cdot f_{k-2}^l + 2 \cdot f_{k-2}^{l-1} + 3 \cdot f_{k-2}^{l-2} + \ldots + l \cdot f_{k-2}^1, \quad l \geqslant 1, k \geqslant 2,$$

$$(8.23)$$

(in modern notation, f_k^l being the *l*th figurate number in k
dimensions, as in Fig. 8). Thus the triangular numbers are
successively $1 \cdot 1 = 1$, $1 \cdot 1 + 2 \cdot 1 = 3$, $1 \cdot 1 + 2 \cdot 1 + 3 \cdot 1 = 6$, and
the tetrahedral numbers successively $1 \cdot 1 = 1$, $1 \cdot 2 + 2 \cdot 1 = 4$,
$1 \cdot 3 + 2 \cdot 2 + 3 \cdot 1 = 10$, and so on. Leibniz observed the relation for
the tetrahedral numbers by means of a diagram of dots similar in
form to Fig. 12: note how easy it is to see that the tetrahedral
number $35 = 1 \cdot 5 + 2 \cdot 4 + 3 \cdot 3 + 4 \cdot 2 + 5 \cdot 1$. In fact (8.23) is most
naturally thought of as arising from a double application of

$$f_k^l = \sum_{i=1}^{l} f_{k-1}^i. \qquad (1.7 \; bis)$$

A total of $(s + 1)$ applications of (1.7) leads to the expansion of a
figurate number f_k^l in terms of the inner product of the first l
number $35 = 1 \cdot 5 + 2 \cdot 4 + 3 \cdot 3 + 4 \cdot 2 + 5 \cdot 1$. In fact (8.23) is most
naturally thought of as arising from a double application of

$$f_k^l = f_s^1 \cdot f_{k-s-1}^l + f_s^2 \cdot f_{k-s-1}^{l-1} + f_s^3 \cdot f_{k-s-1}^{l-2} + \ldots + f_s^l \cdot f_{k-s-1}^1,$$

$$l \geqslant 1, \quad s \geqslant 0, \quad k \geqslant s + 1. \quad (8.24)$$

Leibniz seems not to have known this full expansion [41], however,
nor the Euler expansion derived by repeated application of

$$f_k^l = f_k^{l-1} + f_{k-1}^l, \qquad (1.6 \; bis)$$

though this is sometimes attributed to him [42].

Notes to Chapter 8

[1] Accounts of Wallis's work may be found in Scott (1938) *The Mathematical Work of John Wallis* and, with special reference to the *Arithmetica infinitorum*, in the papers by Nunn (1910) and Whiteside (1961a) and chapter 7 of the book by Edwards (1979), pp. 166–76. Some of Wallis's Propositions from the relevant part of *Arithmetica infinitorum* are translated in Struik (1969), pp. 244–53.

[2] Wallis (1655).

[3] See Chapter 1.

[4] Wallis (1655), p. 392.

[5] Whiteside (1961a), p. 237.

[6] Wallis (1655), p. 423.

[7] Wallis (1655), p. 424.

[8] Wallis (1655), p. 424.

[9] Wallis (1655), pp. 443–8.

[10] Whiteside (1961a), p. 236: "Few 17th century mathematicians seem to have read Briggs' lengthy and apparently obscure introductions to his tables—certainly not Wallis".

[11] Wallis (1655), p. 458.

[12] Whiteside (1961a), p. 239.

[13] Wallis (1655), p. 461.

[14] Wallis (1655), p. 462.

[15] Wallis (1655), p. 467.

[16] Edwards (1979), p. 175, gives the impression that Wallis used the monotonicity of the successive values directly in order to obtain his result, but this is not correct; that suggestion was made by Newton in his annotation of Wallis (Newton, 1664, pp. 103–4).

[17] Wallis (1655), p. 459, Proposition 184: "Moreover, all that has been said in the preceding propositions about the interpolation of one number in each interval can be easily extended to the interpolation of two, three, or more numbers in each interval, *mutatis mutandis*".

[18] Youschkevitch (1971), p. 478.

[19] The *Epistola posterior of* 24 October 1676: Newton (1676), pp. 110–12 (Latin) and 130–32 (English); also translated in Smith (1929), pp. 225–8 and Struik (1969), pp. 287–90. Modern accounts of Newton's discovery of the general binomial theorem (for example Edwards, 1979, pp. 178–88) draw their inspiration from Whiteside (1961b); but Whiteside was quite explicitly describing Newton's revised version (Autumn 1665?) in Cambridge University Library manuscript Add 3958.3:70–71 rather than the earlier Add 4000:18–19v from the winter of 1664–5 (Whiteside's edition of these manuscripts is Newton (1664, pp. 104–34). Our account is based on the earlier manuscript. After several months' reflection on this manuscript (both in its original and published forms) I came to believe that Newton had discovered the theorem by making a very direct extension indeed of one of Wallis's tables. Only after reaching this conclusion did I consult J. M. Child's "Newton and the art of discovery" (Greenstreet, 1927, pp. 117–29), and it has been gratifying to learn that Child's conclusion was identical. This is all the more remarkable since Child did not know of Add 4000. See also Whiteside's comments in Newton (1674), pp. 666–74.

[20] Newton (1664), p. 96; see also the above note.

[21] Newton (1664), p. 107.

[22] In later life Newton became familiar with Pascal's *Traité* itself: see Hofmann (1974), p. 29 note 28; Collins (1712), p. 37.

[23] Wallis (1655), p. 459: Proposition 184.

[24] I hope to give my detailed reasons for this conclusion in a separate publication.

[25] It so happens that this rule is not given in any of the books which Newton is known to have read (for which see Whiteside, 1967, pp. 19–24).

[26] The *Epistola prior* of 13 June 1676: Cambridge University Library manuscript Add 3977.2; Newton (1676), pp. 20–32 (Latin) and 32–41 (English); also translated and given in part in Smith (1929), pp. 224–5 and Struik (1969), pp. 285–7.

[27] Wallis (1685a), pp. 318–20; chapter LXXXV.

[28] Or rather, misquotes it, for he omits the term $(m - 4)/5$.

[29] Whiteside (1972), p. 528; Edwards (1979), p. 284.

[30] Euler (1775); Boyer (1968), p. 555.

[31] Hofmann (1974), p. 3.

[32] Leibniz (1666). Accounts of it are to be found in Todhunter (1865), pp. 31–3, and Maistrov (1974), pp. 36–8, whilst Knobloch (1973), pp. 23–53, gives an exhaustive analysis; see also Hofmann (1974), pp. 3–5.

[33] Schwenter (1636); see also Hofmann (1974), pp. 4–5.

[34] See our Chapter 4. Leibniz gave a proof of the general result in 1681 (Dickson, 1919, Volume I, pp. 59–60). If n is prime, and $r < n$, n and $r!$ have no common factor, so that

$$\binom{n}{r} = n \cdot \frac{(n - 1)(n - 2)(n - 3) \ldots (n - r + 1)}{r!}, \quad 1 \leqslant r \leqslant (n - 1),$$

where the second part of the product must be an integer. The property is important in the proof of Fermat's theorem (Dickson, 1919, Vol. I, pp. 59–60; Hardy and Wright, 1954, pp. 63–6).

[35] Our account relies principally on Hofmann (1974), but see also Edwards (1979), pp. 231–67, and Boyer (1968), pp. 438–43.

[36] Hofmann (1974), p. 20; also given by Leibniz in his letter to Oldenburg dated 3 February 1673 (Collins, 1712, p. 36). Neither Boyer (1968), p. 439, nor Edwards (1979), p. 237, give this table, but only the related one (our Fig. 31) needed for clarity of exposition.

[37] Boyer (1968), p. 293; Struik (1969), p. 320, translates Oresme's demonstration.

[38] Natucci (1974), p. 303; the sums were published in Pietro Mengoli's *Novae quadraturae arithmeticae*, Bologna, 1650.

[39] For an account, see Knobloch (1973), pp. 59–90; Leibniz's combinatorial manuscripts have been published by Knobloch (1976).

[40] Knobloch (1976), pp. 10–15; Knobloch (1973), pp. 60–3.

[41] Needham (1959), pp. 138–9, apparently quoting Sarton (1947), Part I, p. 702, suggests that (8.24) was known to Chu Shih-chieh in 1303, but here the caution I expressed in note 50 to Chapter 1 seems in order. The attribution is in any case queried by Ho Peng-Yoke (1971), pp. 268–9. Knuth (1973), p. 58, apparently misreading Needham, attributes Euler's identity (for which see note 42) to Chu Shih-chieh.

[42] Maistrov (1974), pp. 37–8, says that Leibniz's later works include "results for the number of outcomes (without repetitions) with 1, 2, 3, etc., dice [which] can be expressed in the following manner:

$$1 \text{ die } \quad \binom{6}{1} = 6,$$

$$2 \text{ dice } \quad \binom{1}{0}\binom{6}{1} + \binom{1}{1}\binom{6}{2} = \binom{7}{2} = 21,$$

$$3 \text{ dice } \quad \binom{2}{0}\binom{6}{1} + \binom{2}{1}\binom{6}{2} + \binom{2}{2}\binom{6}{3} = \binom{8}{3} = 56,$$

and so on up to six dice", which is the correct solution to Tartaglia's problem (see our Chapter 4). Knobloch (1973), p. 69, gives a similar description. However, this must not be taken to imply that Leibniz possessed the Euler expansion of a binomial coefficient

$$\sum_k \binom{r}{k}\binom{s}{n-k} = \binom{r+s}{n}, \quad n \leqslant r + s; \; k = 0, 1, 2, 3, \ldots,$$

with the convention that $\quad \binom{a}{b} = \binom{a}{a-b} = 0$ for $b > a$,

which arises from repeated application of the addition rule of formation (5.1) or from considering the coefficient of x^n in the identity $(x + r)^{r+s} = (x + y)^r (x + y)^s$. In order to see this point we must examine Leibniz's solution for four dice contained in his note *Ex doctrina combinationum* (some time after January 1708; Knobloch, 1976, pp. 224–6). Working with the letters a, b, c, d, e, f rather than the points of a die, he calculates as follows (though not in this tabular form):

Type of selection	Number of such types	Selections in each type		Total
a^4 (i.e. all four the same)	1	×	$\binom{6}{1}$ =	6
a^3b	2 (i.e. a^3b, ab^3)	×	$\binom{6}{2}$ =	30
a^2b^2	1	×	$\binom{6}{2}$ =	15
a^2bc	3	×	$\binom{6}{3}$ =	60
$abcd$	1	×	$\binom{6}{4}$ =	15
				126

He notes that $126 = (6 \cdot 7 \cdot 8 \cdot 9)/(1 \cdot 2 \cdot 3 \cdot 4)$ (which we may write as $\binom{9}{4}$), but he does *not* recognize that the total number of types involving just two kinds of letter (rows 2 and 3 above) is the binomial coefficient $\binom{3}{1} = 3$. Rather, he enumerates the numbers of each type separately, using the multinomial coefficient since the problem is then that of the number of arrangements of the *indices*. Thus for a^2bc he finds "$(1 \cdot 2 \cdot 3)/(1 \cdot 2; 1) = 3$" such types, because there are three indices $(2, 1, 1)$, two of one kind and one of another. Thirty-two years earlier Leibniz had forgotten to include the type a^2b^2 altogether (Knobloch, 1973, pp. 69–70).

We may note that in order to find the total number of types involving exactly l kinds of letter distributed in m indistinguishable places we would have to solve the primary ("dice") problem anyway, so that Leibniz's supposed solution contains a circularity. For l of the places can each be assigned a different letter, leaving $m - l$ places to be filled by choosing without restriction from l kinds of letter, which is the problem of the number of throws that can be obtained with $m - l$ l-sided dice

$$= \binom{m - l + l - 1}{m - l} = \binom{m - 1}{l - 1}, \text{ the binomial coefficient required}$$

This provides the combinatorial explanation for the particular case of the Euler expansion involved.

Euler gave the expansion in the paper in which he proved the binomial theorem for all rational indices (Euler, 1775), and repeated it as a theorem a few years later (Euler, 1784).

9

The binomial and multinomial distributions

Nowadays, perhaps the commonest occurrence of the Arithmetical Triangle is in connection with the statistical distribution known as the *binomial*. Let there be n independent trials each resulting in either a *success* (with probability p) or a *failure* (with probability $1 - p$), and let X be the number of successes. Then the probability that the random variable X takes the particular value r is

$$P(X = r) = \binom{n}{r} p^r (1 - p)^{n-r}, \quad r = 0, 1, 2, \ldots, n. \quad (9.1)$$

We saw in Chapter 7 that Pascal evidently possessed this form for $p = \frac{1}{2}$ in 1654, though we must allow for the fact that he thought in terms of *expectations* rather than *probabilities*, so that (9.1) would have been the proportion of the total stake which it would have been reasonable to place on the result $X = r$. We also saw that his proof did not rely on combinatorial arguments, and that he obtained the coefficients directly from the Arithmetical Triangle.

The distribution takes its name, of course, from the fact that its successive terms are those of the expansion of the binomial $(p + q)^n$ where $q = 1 - p$; in modern terminology we say that $(pz + q)^n$ is the *generating function* of the distribution, z being a dummy variable. The fact that the coefficients of the terms of the distribution can be viewed equally as combinatorial or binomial in origin gives the distribution a rather special place in the post-Pascalian history of the Arithmetical Triangle, and it is a happy fact that we may trace its beginnings to Pascal himself.

It is now common to introduce the binomial distribution by arguing (as did Thomas Bayes in his famous *Essay* [1], for example) that any particular sequence of r successes and $n - r$ failures has probability $p^r(1 - p)^{n-r}$, and that there will be $\binom{n}{r}$ such sequences because this is the *number of arrangements of r things of one kind and $n - r$ of another*. Pascal, as we have seen, probably never thought in terms of numbers of arrangements, but at some point in the late seventeenth century writers in the West began to appreciate the isomorphism between the problems of finding the number of combinations of n different things taken r at a time and the number of arrangements of r things of one kind and $n - r$ of another, an isomorphism already recognized by the Hindu mathematician Bhaskara in twelfth-century India (see our Chapter 3). The simplest explanation is to imagine the n different things to be in a particular order and then to call the r that have been selected in the first problem "things of the first kind" and the $n - r$ left behind "things of the second kind", whereupon the second problem appears, but in fact it seems that the result was obtained by applying the general multinomial rule to the case of alternatives and recognizing the result as ^nC_r.

This rule, that the number of arrangements of a things of one kind, b of another, c of another, and so on, is equal to

$$\frac{n!}{a!\,b!\,c!\,\dots}, \qquad (3.3\ bis)$$

first appears in the West in the work of Mersenne in 1636 and was explained by Wallis in 1685 (see our Chapter 4); Bhaskara had already given it in the East. n different things can be arranged in $n!$ ways, but if any a of them (which can be arranged amongst themselves in $a!$ ways) should be identical the number of arrangements is thereby reduced to $n!/a!$, and so on for $b, c, \dots$.

On 16 May 1695 Leibniz [2] wrote to John Bernoulli "I have conceived then of a wonderful rule for the coefficients of the powers not only of the binomial $(x + y)$, but also of the trinomial $(x + y + z)$, in fact, of any polynomial; so that when given the power of any degree say the tenth, and any term contained in it $\dots$, it should be possible to assign the coefficient $\dots$". John Bernoulli replied on 8 June [3] giving the multinomial formula (3.3 bis) and adding "It would be a pleasure to see your rule and it would be well to test whether they agree; yours is possibly simpler". We know from Leibniz's notes that his form of the rule was, for a things of one kind, b of another, c of another, ..., f, there being n things all told,

$$\frac{n(n-1)(n-2)\ldots(f+1)}{a!\,b!\,c!\,\ldots\,e!} \tag{9.2}$$

which is of course the same as Bernoulli's form

$$\frac{n!}{a!\,b!\,c!\,\ldots\,e!\,f!}. \tag{3.3 bis}$$

Leibniz derived (9.2) from

$$^{n}C_{a}\cdot{}^{n-a}C_{b}\cdot{}^{n-a-b}C_{c}\ldots{}^{n-a-b\ldots-e}C_{f}, \tag{9.3}$$

though without adding Montmort's later explanation that the first factor gives the number of ways a similar things can be distributed in n places, b things in the remaining $(n-a)$ places, and so forth, which he surely understood [4]. It is interesting to find, so early in the life of the multinomial coefficients, a form which demonstrates that each is the product of binomial coefficients. Leibniz seems to have known (9.3) at the beginning of the 1680's [5], and possibly obtained it from Kircherus (for whom see Chapter 4), having realized that a multinomial coefficient necessarily records the number of arrangements of the elements of the term. Both Hérigone [6] and Oughtred [7] had expanded the trinomials $(x+y+z)^2$ and $(x+y+z)^3$, but here was an entirely general rule.

   Neither Leibniz's notes nor his correspondence were published at the time, but in England Abraham De Moivre published the multinomial coefficient in 1698 [8], observing simply that (3.3) gives the number of arrangements of the elements ("this is the rule which is commonly given"), which is what the coefficient must be. From the beginning of the eighteenth century, therefore, it was generally known that there was a connection between numbers of arrangements and multinomial coefficients; in *Ars conjectandi* [9] James Bernoulli wrote "It is proper here to note the peculiar *sympathy* between combinations and powers of multinomials" [10]. *Ipso facto* the number of ways of arranging r things of one kind and

$n - r$ of another is the binomial coefficient $\binom{n}{r}$, and thus equal to ${}^nC_r$, a result explicitly given in *Ars conjectandi* [11].

Precisely this connection was made by John Bernoulli in a letter to Montmort dated 17 March 1710 [12] in which he commented on the first (1708) edition of Montmort's *Essay d'analyse sur les jeux de hazard* [13], with the result that in the second (1713) edition Montmort explicitly observed for the binomial what De Moivre had seen generally, that the coefficient of a term in a binomial expansion is given by, and gives, the number of arrangements of its elements [14], precisely because that is how the coefficient can be considered to have been generated, and he then added the multinomial case as well. Indeed, here is the fundamental notion of combinatorial analysis.

Montmort wrote his book knowing of the existence of the manuscript of *Ars conjectandi*, but having despaired of it ever being published. He is familiar with Pascal's *Traité*, with the correspondence between Pascal and Fermat, and with the work of Huygens. In his first edition he gives Pascal's solution to the Problem of Points for two players of equal skill [15] (our 7.5), presumably taken from Part II, Section 3 of the *Traité*, but without proof. The coefficients are given in the form

$$\frac{n(n - 1)(n - 2)\ldots(n - r + 1)}{1.2.3\ldots r}, \qquad (3.1\ bis)$$

by now standard. This provides clear evidence that the notion of the binomial distribution for equal chances was familiar, whilst further evidence, in the form of an anecdote, is provided towards the end of the book. Montmort is considering the Game of Nuts, apparently played in Canada, in which eight nuts, each with one black and one white side, are tossed, and "If the number of black is odd, he who has thrown the nuts wins the other gambler's stake. If they are all black or all white he wins double stakes, and outside these two cases he loses his stake" [16]. Montmort goes on: "I think I should add that this Problem was proposed to me by a Lady, who gave me almost immediately the correct solution using the Arithmetical Triangle" [17].

Another example of the use of the binomial distribution for equal chances at about this time is afforded by Arbuthnott's celebrated *An*

Argument for Divine Providence, taken from the constant Regularity observ'd in the Births of both Sexes [18], presented in 1710 and published in 1712, in which we read "If the Number of Dice of Two sides, *M* and *F* be *n*, all their Chances will be expressed in this Series

$$M^n + \frac{n}{1} \times M^{n-1}F + \frac{n}{1} \times \frac{n-1}{2} \times M^{n-2}F^2$$

$$+ \frac{n}{1} \times \frac{n-1}{2} \times \frac{n-2}{3} \times M^{n-3}F^3 +, \text{ etc.}"$$

In this form *M* and *F* are dummy variables, not probabilities: "Let the Binome *M* + *F* be raised to the Power, whose Exponent is the Number of Dice given; the Coefficients of the Terms will show all the Chances sought". Thus the binomial generating function, implicit in the work of Pascal, is here used explicitly.

In this first edition of his *Essay* Montmort gives a new demonstration of the combinatorial formula

$$^nC_r = \frac{n(n-1)(n-2)\dots(n-r+1)}{1.2.3\dots r}, \qquad (3.1 \ bis)$$

one which is now quite familiar. He points out [19] that the first thing can be chosen in *n* ways, leaving (*n* − 1) behind, so that the second can be chosen in (*n* − 1) ways, and so on, giving the numerator as the number of ordered selections of *r* things from *n*; but each such ordered selection is one of the *r*! arrangements of that selection, whence the result. It is very close to the argument which we suggested, in Chapter 3, might have been used to derive the rule in India.

Montmort also gives the multinomial coefficient in the form of (9.3) in connection with throws of dice, though without any proof [20]; this is the first publication of a multinomial distribution for equal chances. On seeing it, John Bernoulli, in his letter of 17 March 1710 referred to above, told Montmort of the form of the multinomial coefficient (3.3) [21], and Montmort remarked (in a letter to Nicholas Bernoulli dated 8 June 1712) "It is indeed very curious to see two such different Problems united under the same formula" [22], and proceeded to give his explanation for the fact, which was, of course, the same as De Moivre's; but he added the observation that one can get from the binomial $(x + y)^n$ to the trinomial $(x + y + z)^n$ by replacing y by $(y + z)$ in the expansion of the former, and then one can proceed to the quadrinomial, and so on. In other words, one sets $(x + y + z \ldots + u + v)^n = (x + (y + (z + \ldots (u + v) \ldots)))^n$ and repeatedly applies the binomial expansion to the second expression, a method already given by the Arabic writer Al-Zanjani in 1297 [23]. As a result of his correspondence with the Bernoullis, Montmort's second edition deals more fully with the multinomial coefficient [24]; he explains the form (9.3), makes use of (3.3) in passing, and notes the *rapport* ("curious enough, it seems to me") between the coefficients of a multinomial expansion, the probabilities of particular dice throws, and the numbers of arrangements of things of various kinds.

Another contribution which John Bernoulli made in his letter of 17 March 1710 was the generalization of Pascal's solution to the Problem of Points (7.5) to accommodate two players of unequal skill [25]. As with the special case, this involves summing terms in a binomial distribution, which Montmort therefore gives completely generally [26] in his second edition; Todhunter gives a good account [27]. The reader who refers to Montmort's original should not be dismayed to find the solution there described as having been obtained "by Fermat's method"; Montmort has just reprinted the Fermat–Pascal correspondence, and, as we did in Chapter 7 [28], refers to the solution by enumeration as Fermat's method, in contrast to Pascal's method of taking expectations. But the full binomial solution was, as we saw, obtained by Pascal in Section 3 of Part II of *The Arithmetical Triangle*, and it is the generalization of this that Montmort gives.

De Moivre, also, had generalized the Problem of Points, and thus given the general binomial distribution, in the first edition of his *De mensura sortis* [29] published in 1712, though without Montmort's explanation of its relevance; but he went on [30] to give the multinomial analogue for more than two players, sorting out the terms of the expansion according to which player would have won the corresponding game, not forgetting that some terms had to be split because the winner was determined by the *order* of the throws. Since Pascal and Fermat had proceeded, in the case of three players, by enumerating all the 3^3 possible games, they cannot be said to have used the multinomial distribution for equal chances.

We may note in passing that Montmort also remarked to Nicholas Bernoulli [31] in his letter of 8 June 1712 that he had discovered what no-one he knew had yet found, that the number of terms in a q-nomial raised to the pth power is given by the figurate number f_{q-1}^{p+1}. He gave no explanation, perhaps regarding the result as obvious. Consider, for example, systematically displaying the terms of a trinomial expansion; they fall conveniently onto a triangular lattice (Fig. 32) and their number is thus given by a triangular number. It is the same problem as finding the number of ways p q-sided dice can fall, solved for $q = 6$ by Tartaglia in 1523 (see our Chapter 4), and is given as a corollary to this in Montmort's second edition [32]. However, the result will not have been new to Nicholas Bernoulli because, as we see in our next chapter, it is given in chapter VIII of Part II of his uncle's *Ars conjectandi* (although Nicholas did not mention this when replying to Montmort on 11 October 1712 [33]).

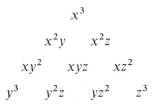

Fig. 32   The terms of $(x + y + z)^3$ displayed to show the connection with the triangular number 10

Summarizing our knowledge so far of the origins of the binomial and multinomial distributions, the binomial with equal chances first appeared in print in all but name in Section 3 of Part II of Pascal's *Treatise*, published in 1665 (written in 1654), whence it found its way into Montmort's first edition, published in 1708, where there is also ample other evidence of its use. The binomial with general chances was published in Montmort's second edition in 1713 (having been suggested to him by John Bernoulli in 1710) and in 1712 in the *Philosophical Transactions* in a paper De Moivre presented in 1710; this paper also contains the multinomial distribution with general chances, the equal-chance version having been used by Montmort in his first edition (1708).

We now turn, in Chapter 10, to the *Ars conjectandi* of James Bernoulli, also published in 1713, where we likewise find the general binomial distribution, probably derived by him as early as 1689, and perhaps the source, via John Bernoulli, of Montmort's knowledge.

Notes to Chapter 9

[1] Bayes (1764), pp. 384–5, Proposition 7. For a commentary see Edwards (1978).
[2] Leibniz (1695), p. 47; in English translation by J. Ginsburg, Smith (1929), p. 229.
[3] Leibniz (1695), pp. 54–5; Smith (1929), p. 230.
[4] Knobloch (1976), p. 81; Montmort (1713), pp. 42–3.
[5] Knobloch (1976), p. 217.
[6] Hérigone (1634), *Algèbre*, p. 19.
[7] Oughtred (1652), p. 62.
[8] De Moivre (1698).
[9] Bernoulli (1713). A description of this work is given in the next chapter. The author was a brother of John Bernoulli, and is called James by English writers, but Jacques in French and Jacobus in Latin.
[10] Bernoulli (1713), p. 131; in English translation by J. Ginsburg, Smith (1929), pp. 230–1.
[11] Bernoulli (1713), p. 103, Corollary 4.
[12] Montmort (1713), p. 286.
[13] Montmort (1708); completed at least a year earlier, since it was registered on 17 May 1707. There were two printings of the second edition (Montmort, 1713), in 1713 and 1714, apparently identical apart from their title-pages: the 1714 printing has the spelling *Essai*. Todhunter (1865), p. 79, refers to the two dates. Both he (pp. 73–134) and Maistrov (1974), pp. 76–80, discuss Montmort's second edition, the former extensively, whilst David (1962), pp. 140–60, concentrates on the first.
[14] Montmort (1713), pp. 32–4.
[15] Montmort (1708), p. 177.
[16] Montmort (1708), pp. 153–5; the translation is from David (1962), p. 149. The problem is repeated in Montmort (1713), pp. 213–15.
[17] David (1962, p. 150) has "this problem was posed by me to a Lady", but the original French certainly indicates the reverse.
[18] Arbuthnott (1712). It seems to me that most commentators have studied Arbuthnott's arguments too superficially, but Hacking (1975), pp. 166–75, is excepted. For Arbuthnott's life see Eisenhart and Birnbaum (1967).
[19] Montmort (1708), pp. 92–4; repeated in Montmort (1713), pp. 22–3.
[20] Montmort (1708), p. 137.
[21] Montmort (1713), pp. 291–2.
[22] Montmort (1713), pp. 353–5.
[23] Yadegari (1980); there is no evidence that this Arabic result influenced the European development described in this chapter.
[24] Montmort (1713), pp. 42–6.

[25] Montmort (1713), pp. 294–5. The suggestion by David (1962), page 157, that it was Nicholas Bernoulli, not John, seems to be a mistake.
[26] Montmort (1713), pp. 244–5.
[27] Todhunter (1865), pp. 96–9.
[28] See especially note 9 of Chapter 7.
[29] De Moivre (1712), p. 217, Problem II. No demonstration or proof is given. An English translation of *De mensura sortis* by B. McClintock is given by Hald (1984), who adds a commentary. For De Moivre's life see Eisenhart and Birnbaum (1967).
[30] De Moivre (1712), pp. 226–7, Problem VIII.
[31] Montmort (1713), p. 355.
[32] Montmort (1713), p. 38.
[33] Montmort (1713), p. 371.

10

Bernoulli's *Ars conjectandi*

Whilst Montmort and De Moivre were publishing their findings, Nicholas Bernoulli was preparing the manuscript of *Ars conjectandi* [1] for the press. Written by his uncle, James Bernoulli, it was published in 1713, eight years after the author's death in 1705. Todhunter [2] and Maistrov [3] describe it, whilst David [4] gives some background material.

This famous work is chiefly noted for the first limit theorem in probability, "Bernoulli's theorem", which is to be found in Part IV; but even without this it would be of great significance because of its treatment of the binomial distribution, the theory of combinations, and the figurate numbers (especially their use in the summation of powers, which led to the discovery of the "Bernoulli numbers" of analysis). Indeed, we shall not mention Part IV again, for though it concerns the future history of the binomial distribution, it is soon time to let that fledgling distribution depart, and to return to our own study of the Arithmetical Triangle. Apart from the accounts of Part IV given by the authors cited above, Hacking [5] and Shafer [6] have more recently analysed its content. Parts I and II contain so much relevant material that it is best to describe them in order rather than to remark on individual points of interest, as was possible with the work of Montmort and De Moivre.

Part I of *Ars conjectandi* is a commentary, with text, on Huygens's book *De ratiociniis in aleae ludo* which had appeared [7] in 1657. In spite of its importance in the early history of probability, we have hardly noted Huygens's work because it contains little of direct relevance to the Arithmetical Triangle, though Huygens did give Pascal's "expectation" solution to the Problem of Points [8]. The standard authors [9] describe his contribution.

First Bernoulli gives Huygens's Propositions I–IX concerning the Problem of Points, with annotations; after Proposition VII he has added a table for the division of stakes between two players whose derivation he is to give in Part II, whilst the table for three players given after Proposition IX is Huygens's own. In Propositions X–XIV dice throws are considered; after his annotation on Proposition XII Bernoulli devotes a section [10] to developing the binomial distribution for general chances, finding the expression for obtaining at least m successes in n trials. This, of course, involves giving the sum of $(m - 1)$ terms of the distribution, which Bernoulli does explicitly. Todhunter says "This formula involves a solution of the Problem of Points for two players of unequal skill; but James Bernoulli does not explicitly make the application" [11].

Bernoulli's derivation of the binomial distribution is the modern one we gave at the beginning of the last chapter except that his argument for the coefficients being given by ^nC_r is direct: consider, he says, n dice labelled A, B, C, D, ...; then if r of them show a "success" those r can be chosen from the n different dice in ^nC_r ways [12]. As we saw then, this argument was not the one used by Montmort, who had originally obtained the equal-chances binomial distribution from Pascal and then added the "number of arrangements" argument drawn to his attention by John Bernoulli. Since we shall shortly provide evidence that although James Bernoulli was familiar with Pascal's letters to Fermat he was unaware of Pascal's *Treatise*, it seems very likely that his derivation of the binomial distribution was not influenced by any of the other derivations. We do not know in what year he obtained it, but it was probably between 1685 and 1689 [13]. Then all three accounts of the general binomial, Bernoulli's, Montmort's, and De Moivre's, appeared virtually simultaneously, in 1712 and 1713.

Huygen's book ended with five problems for solution [14]; Todhunter [15] describes Bernoulli's contributions, which need not concern us.

Part II of *Ars conjectandi* is *The Doctrine of Permutations and Combinations*, and shows that Bernoulli was not familiar with Pascal's *Treatise*. Although he refers to the 1679 edition of Fermat's correspondence [16], he only mentions Van Schooten, Leibniz, Wallis and Prestet as his predecessors [17] in combinatorial theory; and later, when discussing the figurate numbers [18], he adds the names of Faulhaber, Remmelin, and Mercator [19], but never Pascal. Moreover, his is ignorant of Pascal's proof of the famous Corollary 12 (6.12) of the *Treatise*, saying, in the Preface to Part II, "I have added certain material [to the work of my predecessors] which is not to be despised, especially the general and straightforward proof of that property of the figurate numbers on which a large part of the rest relies, [a proof] which, so far as I know, has never been given or brought to light before" [20]. His proof [21] turns out to be longer and less elegant than Pascal's, though it, too, proceeds by induction. "At the time when I showed this proof to my brother, he pointed out that it could be elegantly shortened" [22], and James adds the shortened version to his text, but it is still not Pascal's. It is thus clear that neither John nor James Bernoulli knew of Pascal's *Treatise* at the time. Just before he died in 1705 James's attention was drawn to it by Leibniz [23], but John was still unfamiliar with it in 1710 [24].

Chapter I of Part II concerns permutations, and gives the usual "$n!$" and "$n!/(a!\,b!\,c!\ldots)$" rules; chapter II is on the combinatorial rules "$2^n - 1$" and "$2^n - n - 1$". We have discussed all these in our earlier chapters. Bernoulli's chapter III is on combinations of different things taken 1, 2, 3, ... at a time, and on the figurate numbers, "by which these matters may be treated". It opens with the statement that, following on from the pattern of enumeration he used in the preceding chapter to derive the "$2^n - 1$" rule, namely

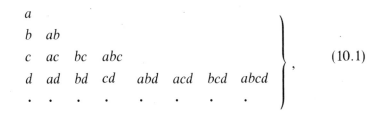

$$(10.1)$$

it is obvious that there is one pair (*ab*) in the second row, two in the third (*ac, bc*), three in the fourth (*ad, bd, cd*), and so on, since the pairs in each row are formed by adding the initial letter of the row to the single letters of the preceding rows. Similarly, the triplets are formed by adding the initial letter of the row to the pairs of the preceding rows, and so on for the higher-order combinations. We may bring this pattern into sharper relief by parsing (10.1) thus:

$$
\begin{array}{lllll}
a \\
b & ab \\
c & (ac,\ bc) & abc \\
d & (ad,\ bd,\ cd) & (abd,\ acd,\ bcd) & abcd \\
e & (4\ \text{pairs}) & (6\ \text{triplets}) & (4\ \text{quadruplets}) & abcde \\
\cdot & \cdot & \cdot & \cdot & \cdot
\end{array} \Biggr\} \ (10.2)
$$

It is now quite clear that the numbers of singles (first column) form the series $(1, 1, 1, 1, \ldots)$, the numbers of pairs (second column) form the series $(1, 2, 3, 4, \ldots)$ by summation from the series for singles, the numbers of triplets (third column) form the series $(1, 3, 6, 10, \ldots)$ by summation from the series for doubles, the successive series being, of course, the figurate numbers, as Bernoulli noted when he constructed his table of combinations (Fig. 33).

Tabula

Combinationum, seu Numerorum Figuratorum.

Exponentes Combinationum.

	I.	II.	III.	IV.	V.	VI.	VII.	VIII.	IX.	X.	XI.	XII.
1.	1	0	0	0	0	0	0	0	0	0	0	0
2.	1	1	0	0	0	0	0	0	0	0	0	0
3.	1	2	1	0	0	0	0	0	0	0	0	0
4.	1	3	3	1	0	0	0	0	0	0	0	0
5.	1	4	6	4	1	0	0	0	0	0	0	0
6.	1	5	10	10	5	1	0	0	0	0	0	0
7.	1	6	15	20	15	6	1	0	0	0	0	0
8.	1	7	21	35	35	21	7	1	0	0	0	0
9.	1	8	28	56	70	56	28	8	1	0	0	0
10.	1	9	36	84	126	126	84	36	9	1	0	0
11.	1	10	45	120	210	252	210	120	45	10	1	0
12.	1	11	55	165	330	462	462	330	165	55	11	1

Numeri Rerum Combinandarum.

Fig. 33 The Combinatorial Triangle from James Bernoulli's *Ars conjectandi* of 1713

Note that the table as it stands does not give ^nC_r in the nth row and rth column: that place, as may be seen from (10.2), gives the number of combinations of order r ("*exponens*" is the word used by Bernoulli, who borrowed it, and this form of the Arithmetical Triangle, from Leibniz's *Ars combinatoria* [25]) which may be formed from n things *and which include the last thing*. In order to find ^nC_r we must form the sum as indicated by Bernoulli, and therefore look in row $(n + 1)$ and column $(r + 1)$. The essence of Bernoulli's method is

$$^nC_r = \sum_{i=r-1}^{n-1} {}^iC_{r-1} \quad \text{with } {}^nC_0 = 1 \text{ for all } n. \tag{10.3}$$

This argument by enumeration will probably have been suggested to Bernoulli by Leibniz's table explaining how the ten 3-combinations of five things can be derived from the 2-combinations of two, three, and four things [26], and it is a pity it has been forgotten because it makes an excellent teaching method. Bernoulli's order of enumeration (10.1) is much more revealing than Van Schooten's (2.1) which is subtly different [27].

Bernoulli was immediately aware, of course, that he was dealing with the figurate numbers: (10.3) is just (1.7), or Pascal's Corollary 2 (6.2). He waxes quite lyrical about the Arithmetical Triangle: "This Table has truly exceptional and admirable properties; for besides concealing within itself the mysteries of combinations, as we have seen, it is known by those expert in the higher parts of mathematics also to hold the foremost secrets of the whole of the rest of the subject" [28]. He then lists twelve "wonderful properties" of the Triangle, rather as Pascal had done, and concludes with the proof of the multiplicative rule of formation (3.1) which, as we have already noted, he thought original. It is interesting that just as Pascal had failed to find a combinatorial explanation for the rule so, I suppose, did Bernoulli, for surely he would have given one had he known it, instead of his lengthy direct proof.

Immediately after proving the multiplicative rule Bernoulli, in a *scholium* [29], uses it in the summation of the powers of the integers. As we saw in Chapter 1, first Faulhaber and then Fermat had already discovered the rule. Femat was content to imply, in 1636, that the sum of any powers could be found by using it, as indeed they could in principle, but he only gave the sum of the fourth powers explicitly [30]. However, it is one thing to observe the existence of a method which can be used to deliver a formula for the sum of the rth powers, and another thing to discover the working rule by which such formulae may be constructed in practice.

Perhaps Pascal had tried and failed, and this is what led him to give the simpler rule described in Chapter 7 (equation 7.12), but he did not achieve the general solution (7.19) in which the sums are expressed as simple polynomials in n. The remarkable Faulhaber had discovered how to construct the polynomials by a rule which "is infinite and can be continued quite generally as far as one wishes" [31]; using his rule he had given the polynomials up to $\sum n^{17}$ by 1631*, but it did not enable him to write down their coefficients directly. Rather, he had solved the seemingly more complex problem of finding the polynomials in $n(n + 1)$, whence he was able, by expansion, to obtain the polynomial in n for each particular case [32].

Bernoulli, by contrast, discovered the rule for finding the coefficients that occur in (7.19) directly, but he gave neither proof nor explanation. He opens his *scholium* by mentioning Faulhaber as one of the people who have studied the figurate numbers, and then, using the multiplicative rule (1.18), shows explicitly how to use it to find the sums up to $\sum n^3$, just as Fermat had done. He observes that the results for higher powers can similarly be found with a little effort and gives a table *Summae Potestatum* (our Fig. 34) which is simply (7.19) written out in full for exponents $r = 1, 2, 3, \ldots, 10$, but with the terms in each polynomial in descending order. (The table contains an error which we mention below.)

Bernoulli then announces the rule for finding the coefficients directly, without the laborious business of recurrent substitution. Smith, writing in 1929, says "Mathematicians have not been able as yet to find by what process Bernoulli derived the properties of his numbers" [33], yet to anyone steeped in the figurate numbers, as was Bernoulli at the time, the pattern is obvious enough, and Bernoulli derived no other "properties". "Indeed", says Bernoulli [34], "anyone who carefully inspects the rule of progression in the

---

* We continue to use the notation of equation 7.12.

Summæ Poteſtatum.

$$\int n \ \infty\ \tfrac{1}{2}nn + \tfrac{1}{2}n.$$
$$\int nn \ \infty\ \tfrac{1}{3}n^3 + \tfrac{1}{2}nn + \tfrac{1}{6}n.$$
$$\int n^3 \ \infty\ \tfrac{1}{4}n^4 + \tfrac{1}{2}n^3 + \tfrac{1}{4}nn.$$
$$\int n^4 \ \infty\ \tfrac{1}{5}n^5 + \tfrac{1}{2}n^4 + \tfrac{1}{3}n^3 \ \ast - \tfrac{1}{30}n.$$
$$\int n^5 \ \infty\ \tfrac{1}{6}n^6 + \tfrac{1}{2}n^5 + \tfrac{5}{12}n^4 \ \ast - \tfrac{1}{12}nn.$$
$$\int n^6 \ \infty\ \tfrac{1}{7}n^7 + \tfrac{1}{2}n^6 + \tfrac{1}{2}n^5 \ \ast - \tfrac{1}{6}n^3 \ \ast + \tfrac{1}{42}n.$$
$$\int n^7 \ \infty\ \tfrac{1}{8}n^8 + \tfrac{1}{2}n^7 + \tfrac{7}{12}n^6 \ \ast - \tfrac{7}{24}n^4 \ \ast + \tfrac{1}{12}nn.$$
$$\int n^8 \ \infty\ \tfrac{1}{9}n^9 + \tfrac{1}{2}n^8 + \tfrac{2}{3}n^7 \ \ast - \tfrac{7}{15}n^5 \ \ast + \tfrac{2}{9}n^3 \ \ast - \tfrac{1}{30}n.$$
$$\int n^9 \ \infty\ \tfrac{1}{10}n^{10} + \tfrac{1}{2}n^9 + \tfrac{3}{4}n^8 \ \ast - \tfrac{7}{10}n^6 \ \ast + \tfrac{1}{2}n^4 \ \ast - \tfrac{1}{12}nn.$$
$$\int n^{10} \ \infty\ \tfrac{1}{11}n^{11} + \tfrac{1}{2}n^{10} + \tfrac{5}{6}n^9 \ \ast - 1\,n^7 \ \ast + 1\,n^5 \ \ast - \tfrac{1}{2}n^3 \ \ast + \tfrac{5}{66}n.$$

Fig. 34 Bernoulli's expressions for the sums of the powers of the integers. The integral sign is used for summation, the modern sign for proportionality reversed is used for equality, and the asterisks mark places where the polynomials lack terms of particular degrees. Note that the coefficient of n^2 in the sum of the ninth powers is wrong (see my text)

table will be able to extend the table without going through the calculations".

That there *is* a pattern is clear from reading the coefficients columnwise: the coefficient of n^{r+1} is $1/(r+1)$, of n^r is $\tfrac{1}{2}$, and of every other row beginning with n^{r-2} appears to be zero. It is therefore very natural to examine the sequence of numbers in the third column, the coefficients of n^{r-1}:

$$\frac{1}{6}, \frac{1}{4}, \frac{1}{3}, \frac{5}{12}, \frac{1}{2}, \frac{7}{12}, \frac{2}{3}, \frac{3}{4}, \frac{5}{6}.$$

Placing these over the common denominator 12 they are immediately seen to be the natural numbers

$$2, 3, 4, 5, 6, 7, 8, 9, 10,$$

leaving little doubt about how the sequence continues. Similarly, the fourth column gives

$$\frac{1}{30}, \frac{1}{12}, \frac{1}{6}, \frac{7}{24}, \frac{7}{15}, \frac{7}{10}, 1,$$

or, expressed with common denominator 120,

$$4, 10, 20, 35, 56, 84, 120,$$

which Bernoulli will instantly have recognized as the tetrahedral numbers. His growing suspicion that the figurate numbers were involved will have been confirmed as he wrote out the fifth column

$$\frac{1}{42}, \frac{1}{12}, \frac{2}{9}, \frac{1}{2}, 1,$$

to find that they are the sixth-order figurate numbers divided by 252:

$$6, 21, 56, 126, 252.$$

Finally, in his table, the sixth column is

$$\frac{1}{30}, \frac{1}{12}, \frac{1}{2},$$

and here we find that Bernoulli (or his printer) has made a mistake [35], for the second number should be 3/20, whereupon the pattern is maintained in the form

$$8, 36, 120,$$

divided by 240. There is but one number in the seventh column, 5/66.

The initial numbers of the columns from the second onwards were christened the *Bernoulli numbers* by De Moivre in 1730 and Euler [36] in 1755, in ignorance of the fact that by 1631 not only had Faulhaber published them up to

$$B_{16} = \frac{-3617}{6 \times 85}$$

in the same context, but that he possessed a scheme for calculating further numbers. Bernoulli's achievement was to recognize the figurate numbers in the expressions, and thus to isolate the remaining computational difficulty in the numbers at the heads of the columns, which he then observed could easily be computed because the coefficients in each row had to sum to 1 (obviously, since setting $n = 1$ had to give this result, as Faulhaber had realized) so that the first time each Bernoulli number cropped up it could be determined by subtraction. It was Faulhaber's misfortune to have spent many years on a problem the keys to whose solution, the Bernoulli numbers, were not expressible by any simple formula, and it was Bernoulli's good fortune to have isolated these numbers in such a way that their computation was obvious.

The neatest modern algorithm for finding the Bernoulli numbers writes $B_r = (B + 1)^r$ where, after expanding the binomial, the exponents are degraded to subscripts and the equation then used to find B_{r-1} in terms of $B_{r-2}, B_{r-3}, \ldots, B_0$, where B_0 is taken to be 1 and $r \geqslant 2$. Thus with $r = 3$, $B_3 = B_3 + 3B_2 + 3B_1 + B_0$, whence $B_2 = -\frac{1}{3}(3(-\frac{1}{2}) + 1) = \frac{1}{6}$. But this finds the "Bernoulli numbers" which occur in the sums $\sum (n - 1)^r$ and requires $B_1 = -\frac{1}{2}$ rather than $+\frac{1}{2}$. These numbers occur in the first column of the matrix in equation (7.18) and indeed the algorithm immediately follows from the inversion of the matrix in (7.17). For the sums $\sum n^r$ it may be shown by means of a slightly improved formulation [37] that the appropriate algorithm is $B_r = (B - 1)^r$.

We close this account of the discovery of the Bernoulli numbers by noting that although Bernoulli was ignorant of Pascal's *Treatise* he could have known about Pascal's method of finding the sums of the powers (7.12) because it was given in the first edition of Prestet's *Elemens des mathematiques* in 1675 [38] which Bernoulli cited at an earlier point in *Ars conjectandi* [39]. He was therefore in a position to have derived his *Summae Potestatum* by means of an easier recurrence formula than the one he actually used, but there is no suggestion that he did so. On the contrary, he simply seems to have had his detailed study of the figurate numbers rewarded by the gift of a problem known to involve them, and in which they were only just concealed from view. Bernoulli owed nothing to Pascal, but it is a measure of the clarity of Pascal's thought that the best modern presentation of the sums of the powers starts with Pascal's formula, whilst Faulhaber's earlier solution has faded into obscurity [40].

Bernoulli ends his chapter III with an account of the use of the Newton–Gregory formula [41] to sum n terms of a series whose fourth difference is zero, indicating by his treatment how the sum can be obtained in higher-order cases. Montmort [42] discusses the same problem at greater length, giving the inverse formula as well.

In chapter IV Bernoulli introduces the Hindu rule (3.1) for the number of combinations of n things taken r at a time, which he is of course able to do since he proved the multiplicative formula for ^nC_r in his previous chapter; he even works out $^{100}C_{20}$ correctly by way of an example, surely the largest binomial coefficient printed up to that time! He then introduces a number of simple corollaries giving combinatorial interpretations to the properties of the figurate numbers established in chapter III, of which Corollary 4 explicitly identifies ^nC_r as being, in addition, the number of arrangements of r things of one kind and $(n - r)$ of another, whilst Corollary 5 gives the direct argument for

$$^{n+1}C_{r+1} = {}^nC_r + {}^nC_{r+1} \qquad\qquad (7.1 \; bis)$$

which Pascal and Leibniz had given.

In an appendix to this chapter Bernoulli again takes up the question of the Problem of Points (as he promised he would in Part I [43]); Todhunter [44] gives a good summary. Bernoulli's knowledge of the Problem seems to be limited to the contents of Pascal's letter to Fermat of 29 July 1654, which we examined in Chapter 7, and also, of course, to Huygens's Propositions, with which, as we have been, Part I of *Ars conjectandi* was concerned. He says there are two ways of solving the problem: Pascal's method of expectations "which, he said in his letter to Fermat, he was unable to solve", and the combinatorial method "which Pascal seems to have used in his solution of the Problem". The relevant passage in Pascal's letter is "This can be shown, but with a great deal of trouble, by combinatorial methods like those you have worked out, and I have not been able to demonstrate it by this other method which I have just explained to you but only by combinations" [45].

Bernoulli, however, managed to extend "this other method"; in Part I he had given a table for the division of stakes between two players [46], as far as A needing 9 points and B needing 7, and had noticed the construction rule: if, when A lacks a points and B lacks b, A's expectation is written $E(a, b)$, then

$$E(a, b) = \tfrac{1}{2}(E(a - 1, b) + E(a, b - 1)). \qquad\qquad (10.4)$$

He did not use any such notation, nor give any proof, but in Part II he now exhibits [47] the pattern of figurate numbers which lies behind the table, and this enables him to give general expressions and to solve the particular case $a = b - 1$ treated by Pascal in his letter. Bernoulli here contributes nothing new; he is indulging his ability to spot number-patterns based on the figurate numbers, and Pascal had already used (10.4) in obtaining the complete solution in his *Treatise* (see the derivation of our (7.8)). Bernoulli was evidently unaware of this, since he mentions Pascal's inability to complete the solution, and we thus have further strong evidence that he was ignorant of Pascal's *Treatise*.

This is also borne out by the fact that in presenting the second, or combinatorial, method in full generality he does not mention that Pascal has already given the result. As we noted in Chapter 7, Pascal's proof used mathematical induction and the concept of expectation, though almost surely he arrived at the result by the simple combinatorial argument; it is this latter that Bernoulli now presents.

We noted in the previous chapter that John Bernoulli had generalized the Problem of Points to accommodate players of unequal skill; James does not do this in *Ars conjectandi* itself, though in an adjoined tract [48] he applies probability theory to the game of tennis (which is only a glorified Problem of Points!) in the course of which he introduces players of unequal skill. So probably John obtained the idea from his brother.

Chapter V contains another Arithmetical Triangle, this time as the solution to the problem of how many combinations of r things can be made from s different *kinds* of things, repeats being allowed. As with the case of combinations without repeats, Bernoulli solves the problem by skilful enumeration. Let the different kinds of things be $a, b, c, d, \ldots$; these letters themselves indicate the single combinations that may be made, $a, b, c, d, \ldots$. As pairs, first we have aa, then we have those involving the second letter, ab and bb, then those involving the third letter, ac, bc and cc, then those involving the fourth letter, ad, bd, cd, dd, and so on. The triplets are now constructed from the pairs by adding a third letter. First, we add a to the pair aa to form the single triplet made from a alone. Next we add b to the pairs aa, ab, and bb to form the three triplets aab, abb and bbb, and so on. The enumeration is readily displayed systematically:

a	aa	aaa	aaaa	·
b	(ab, bb)	(aab, abb, bbb)	(4 quadruplets)	·
c	(ac, bc, cc)	(aac, abc, bbc, acc, bcc, ccc)	(10 quadruplets)	·
d	(4 pairs)	(10 triplets)	(20 quadruplets)	·
·	·	·	·	·

$$(10.5)$$

Each set of combinations is now seen to be constructed from, and therefore to be as numerous as, all those in the preceding column up to and including those in the same row. Bernoulli gives the corresponding Arithmetical Triangle [49], though this time the format is "rectangular" (cf. (10.5) with (10.2)), and concludes that, in our notation, the figurate number

$$f_k^l = \frac{l(l + 1)(l + 2) \ldots (l + k - 1)}{k(k - 1)(k - 2) \ldots 1} \qquad (1.18 \text{ bis})$$

can equally well be considered as (a) the number of ways of combining $(l - 1)$ things from $(k + 1)$ kinds of things and (b) the number of ways of combining things in combinations of $1, 2, 3, \ldots,$ $(l - 1)$ from k kinds of things, $+ 1$. Of these, (a) arises through taking the column sums of (10.5) and (b) through taking the row sums, in this case the $+ 1$ being necessary because the sum includes the case of selecting nothing. Since by symmetry

$$f_k^l = f_{l-1}^{k+1}, \qquad (6.5 \text{ bis})$$

(a) is more conveniently described as the number of ways of combining k things from l kinds of things. Bernoulli also gives a direct argument, observing that if we *define* f_k^l as the number of ways of combining k things from l different kinds of things,

$$f_k^l = f_k^{l-1} + f_{k-1}^{l-1} + f_{k-2}^{l-1} + \ldots + f_1^{l-1} + 1, \qquad (10.6)$$

since it must equal the number of k-combinations without the lth thing plus the number of $(k - 1)$-combinations without the lth thing plus ... plus the number of l-combinations without the lth thing, plus 1 for the k-combination consisting *only* of the lth thing. In other words, Bernoulli has given a combinatorial interpretation of the column-addition property, demonstrating (a) and (b) simultaneously.

In our Chapter 4 we recorded that in 1523 Tartaglia had solved (a) for the special case of six-sided dice, thus finding the number of ways of combining k things from 6 different kinds of things, and we remarked that there was nothing special about the number 6 in Tartaglia's treatment. Indeed, Tartaglia's Table (our Fig. 14) is simply the first six columns of Bernoulli's. We also allowed ourselves a little speculation on how Tartaglia might have arrived at his solution, and we can now see that Bernoulli has essentially done what we believe Tartaglia did some 150 years earlier. As Bernoulli notes in his chapter VIII, it is also the problem of finding the number of terms in the power of a multinomial (Fig. 32, page 118), and was treated by Montmort in his second edition [50].

With this observation, we take our leave of *Ars conjectandi*. The remaining four chapters of Part II deal with various combinatorial problems which take us further away from the Arithmetical Triangle rather than back to it; we have already noticed Bernoulli's remark (in his chapter VIII [51]) about the "peculiar *sympathy* between combinations and powers of multinomials". Part III is subtitled "The use of the preceding methods in various games of chance and dicing", and Part IV we have already mentioned. Printed together with the book is a memoir on infinite series [52] which discusses, amongst other things, the sums of the reciprocals of the figurate numbers (already known from Leibniz's work [53]).

In his posthumous *Ars conjectandi*, James Bernoulli produced a work which has achieved great fame, and not only for the justifiably famous Part IV. A recent appriasal of Parts I to III is:

"The three parts of Bernoulli's book examined above are of definite interest to the history of mathematics. Some problems of probability theory which by that time had become standard were examined from a new angle. The role of combinatorics in probability theory was fully appreciated. For the first time, the theory of combinations was treated in a systematic manner and many new properties and formulas were obtained. Some very interesting results were also derived in other areas. The first three parts *per se* were thus a significant contribution in the development of mathematics in general and the theory of probability in particular" (Maistrov, 1974 [54]).

But this is surely too high praise. Pascal had treated the theory of combinations half a century before, in a published work of which Bernoulli (unlike Montmort) was ignorant. Which of the "many properties and formulas" were new? From what "new angle" did Bernoulli examine the problems of probability theory? The contribution Bernoulli especially drew to our attention as novel— his proof of the multiplicative property of the figurate numbers— was not as clear as Pascal's; his treatment of the Problem of Points added nothing to Pascal's and his polynomials for the sums of the powers of the integers had been given by Faulhaber. As we have seen, his general binomial distribution was, by virtue of the delay in publication, given later than De Moivre's and contemporaneously with Montmort's (though, as we noted, the latter had it from John Bernoulli, who quite probably obtained it from James).

Todhunter, remarking on Gouraud's (Paris, 1848) comments on *Ars conjectandi*, has this to say:

> "This panegyric, however, seems to neglect the simple fact of the date of *publication* of the *Ars conjectandi*, which was really subsequent to the first appearance of Montmort and De Moivre in this field of mathematical investigation. The researches of James Bernoulli were doubtless the earlier in existence, but they were the later in appearance before the world; and thus the influence which they might have exercised had already been produced. The problems in the first three parts of *Ars conjectandi* cannot be considered equal in importance or difficulty to those which we find investigated by Montmort and De Moivre; but the memorable theorem in the fourth part, which justly bears its author's name, will ensure him a permanent place in the history of the Theory of Probability" [55].

To that assessment we may add two points. First, *Ars conjectandi* is a very well-written book [56] excellently constructed; secondly— and of particular relevance to the present essay—Bernoulli's real *forte* (in Parts I to III) was a thorough familiarity with the many facets of the figurate numbers, so that he could see their patterns shining through the haze which surrounded the various problems.

He picked them out in many combinatorial contexts by clever enumerations, and saw their influence in the Problem of Points; his particular triumph was to see them in the expressions for the sums of the powers of the integers. It seems likely that when Bernoulli finally encountered Pascal's *Traité du triangle arithmétique*, he will indeed have found it of quite extraordinary interest.

Notes to Chapter 10

[1] Bernoulli (1713). The first part of Part I (Propositions I to IX) has been translated into English by H. E. Wedeck and published in Runes (1962), pp. 125–44. Sections of Part II have been published in English as follows: Introduction and chapters I, II and III, with a commentary, by Maseres (1795), pp. 35–213 (it is not possible to distinguish commentary from original in this translation; however, the Latin is given (pp. 1–34) and may be compared); chapter III (pp. 95–8 only), by J. Ginsburg, Smith (1929), pp. 85–90 (reprinted in Struik (1969), pp. 316–20); chapter V, by M. M. Taylor, Smith (1929), pp. 272–7; and the end of chapter VIII (pp. 131–2 only), by J. Ginsburg, Smith (1929), pp. 230–1. Nicholas Bernoulli's Preface is translated in David (1962), pp. 133–5. There is a German translation of the entire work in Ostwald's *Klassiker* (Haussner, 1899).
[2] Todhunter (1865), pp. 56–77.
[3] Maistrov (1974), pp. 56–75.
[4] David (1962), pp. 130–9.
[5] Hacking (1971); Hacking (1975), pp. 143–65.
[6] Shafer (1978).
[7] As an appendix to Van Schooten (1657), of which it is pp. 517–34. The Dutch version, with French translation, is in Huygens (1655), pp. 50–91. There is an early English translation by Arbuthnott (1738) and, according to Todhunter (1865), p. 199, another by W. Browne published in 1714. David (1962), pp. 114–19, carries English translations of the foreword, the propositions, and some of the proofs, but apparently done from the French. See also note 1 above.
[8] Authorities differ over Huygens's source: David (1962), p. 111: "He was not, however, told of the solutions of Fermat and of Pascal nor of the methods which they followed"; Taton (1974), p. 335: "During a stay in Paris in 1655 Huygens was informed in detail of their work": Boyer (1963), p. 290: "The treatise of Huygens was prompted by the Pascal–Fermat correspondence"; Hacking (1975), p. 93: "Nor, it seems, did the young Huygens see any written solution of Méré's problems". Huygens himself wrote (David, 1962, p. 115) "It should be said, also, that for some time some of the best mathematicians of France have occupied themselves with this kind of calculus so that no one should attribute to me the honour of the first invention. This does not belong to me. But these savants, although they put each other to the test by proposing to each other many questions difficult to solve, have hidden their methods. I

have had therefore to examine and to go deeply for myself into this matter by beginning with the elements, and it is impossible for me for this reason to affirm that I have even started from the same principle. But finally I have found that my answers in many cases do not differ from theirs". It is *just* possible to believe that no inkling of Pascal's method of expectations was communicated to Huygens along with the Problem of Points, but difficult. In any event, through correspondence with Carcavi in the autumn of 1656 (Huygens, 1656, pp. 492–4 and 505–7) Huygens learnt that his method was the same as Pascal's, as a result of which he added Proposition IX to his book, giving Pascal's method applied to the Problem of Points succinctly, so that before his book went to press he knew of Pascal's priority in the matter of expectations. Note that in David's account (page 113) she uses the phrase "problem of points" for another of the problems.

  [9] Todhunter (1865), pp. 22–5; David (1962), pp. 110–22; Maistrov (1974), pp. 48–55; Sheynin (1977), pp. 239–41.
 [10] Bernoulli (1713), pp. 38–45.
 [11] Todhunter (1865), p. 60.
 [12] Bernoulli (1713), p. 41; he refers to Part II for the combinatorial argument (which we discuss later in the chapter).
 [13] 1685, when he started working on probability, and 1689, the approximate date of Bernoulli's Theorem (Hofmann, 1970, p. 47).
 [14] The fifth of the problems is the famous "Gambler's Ruin" problem posed by Pascal (see Edwards, 1983); David (1962), p. 118, wrongly describes the first problem as "just the problem of points all over again" (see note 8 above).
 [15] Todhunter (1865), pp. 60–3.
 [16] Bernoulli (1713), pp. 29 and 107. Fermat (1679); the edition contains copies of Pascal's letters to Fermat, but not of Fermat's to Pascal, so that Bernoulli would not have been able to see the reference to Pascal's *Treatise* in Fermat's letter of 29 August 1654.
 [17] Bernoulli (1713), p. 73. Van Schooten (1657) is the book which includes Huygens's work; in Book 5, *Sectiones Miscellaneae, Sectio 1* (pp. 373–80), Van Schooten himself considers some combinatorial questions, giving the enumerative justification for the "$2^n - 1$" rule (see our Chapter 2) and generalizing it to allow repeats. Leibniz (1666); Wallis (1685b); Prestet (1675).
 [18] Bernoulli (1713), p. 95.
 [19] Johann Faulhaber (1615, 1631): see below in this chapter, and also above in Chapter 1. Johann Ludwig Remmelin, of Ulm, whose *Formatio figurati numeri miraculosa* is mentioned by Faulhaber (1631) and given the date 1627 by Kästner (1799), p. 151, but of which I have been unable to find a copy. Nicholas Mercator, who discussed figurate numbers in his *Logarithmotechnia*, London, 1668.
 [20] Bernoulli (1713), p. 73. In the opinion of Professor Brink, the last *which* in this quotation refers to the *proof* and not the *property*. Bernoulli could have learnt of the property from Fermat's letter of 4 November 1636 to de Roberval (see Chapter 1), since this was printed in Fermat (1679), pp. 146–7.
 [21] Bernoulli (1713), pp. 90–4.
 [22] Bernoulli (1713), pp. 92–3.
 [23] Haussner (1899), p. 150, who agrees that Bernoulli was previously unfamiliar with Pascal's *Treatise*.
 [24] For in his letter to Montmort dated 17 March 1710 (Montmort, 1713, p. 290),

commenting on Montmort's first edition, he wrote "It does not seem that Pascal himself understood all the usages of the Table [the Arithmetical Triangle]" and then informed Montmort that it could be used to find the coefficients in the expansion of a binomial!

[25] Leibniz (1666).

[26] See Chapter 8.

[27] Maistrov (1974), p. 61, whilst suggesting that Bernoulli borrowed the pattern from Van Schooten, unfortunately does not record the difference.

[28] Bernoulli (1713), p. 88.

[29] Bernoulli (1713), pp. 95–8; see note 1 above for a translation. Maistrov (1974), pp. 65–6, unaccountably describes the *scholium* out of order, making it appear to be in Bernoulli's chapter IX, which it is not.

[30] Mahoney (1973), p. 230.

[31] Faulhaber (1631), Introduction.

[32] Edwards (1982a); Schneider (1983); Schneider (1982) discusses Faulhaber's influence on Bernoulli.

[33] Smith (1929), p. 85.

[34] Bernoulli (1713), p. 97.

[35] Edwards (1982a); presumably a copying error—cf. lines 5 and 7.

[36] De Moivre (1730), *Supplementum*, p. 19; Euler (1755), p. 335.

[37] Edwards (1982a).

[38] Prestet (1675), pp. 178–81.

[39] Bernoulli (1713), p. 78.

[40] It was rediscovered over two centuries later by Jacobi (1834), who did not mention Faulhaber. However, the copy of Faulhaber (1631) in the Cambridge University Library (Hhh 776) is stamped "C. Jacobi", but there is no indication of whether he acquired it before or after 1834.

[41] Bernoulli (1713), pp. 98–9.

[42] Montmort (1713), pp. 63–72; Todhunter (1865), pp. 85–6.

[43] Bernoulli (1713), p. 16.

[44] Todhunter (1865), p. 66.

[45] Merrington's translation; see Chapter 6, note 7.

[46] Bernoulli (1713), p. 16.

[47] Bernoulli (1713), p. 109.

[48] Bernoulli (1713), *Lettre à un amy, sur les parties du jeu de paume* (35 pages). See also Todhunter (1865), pp. 75–7.

[49] Bernoulli (1713), p. 114. This section has been translated; see note 1 above.

[50] Montmort (1713), pp. 35–8, whose solution is similarly by enumeration. I was not aware of either Montmort's or Bernoulli's solution when I described Tartaglia's, and it has seemed best to leave my description of the latter unchanged, as an independent speculation on how Tartaglia thought.

[51] Bernoulli (1713), p. 131.

[52] Bernoulli (1713), pp. 241–306. Much of this material had been published between 1689 and 1704; part is translated in Struik (1969), pp. 321–4.

[53] See our Chapter 8.

[54] Maistrov (1974), pp. 66–7.

[55] Todhunter (1865), p. 77.

[56] Professor C. O. Brink informs me that even the style of the Latin is excellent.

Appendix 1

Pascal and the Problem of Points

Reprinted from *Int. Statist. Rev.* **50**, 259–66 (1982) by permission of the International Statistical Institute

1 Introduction

When Isaac Todhunter wrote his influential *A History of the Mathematical Theory of Probability* (Todhunter, 1865) he transmitted a tradition that the theory sprang from a correspondence between Blaise Pascal and Pierre de Fermat in 1654 concerning the then-unsolved Problem of Points. Time has treated this tradition well, for in spite of our increased awareness of the earlier work of Cardano (Ore, 1953) and Galileo (David, 1962) it is clear that before Pascal and Fermat no more had been achieved than the enumeration of the fundamental probability set in various games with dice or cards.

An element of the tradition has been that Pascal and Fermat contributed if not equally, then not too disproportionately, to the advances of 1654, but in the last few decades some writers have inclined to the view that Fermat's was the major contribution. Indeed, the basic solution to the Problem of Points is sometimes attributed to him (though, as we shall see, it is Pascal's; see also Ore, 1960), whilst his mathematical biographer (Mahoney, 1973) goes so far as to credit Fermat with quite an important part of Pascal's reasoning. David's popular *Games, Gods and Gambling* (1962) comes down heavily in favour of Fermat:

> "On the whole it would seem that Pascal's reputation as a mathematician is greater than he deserved, ...;"
>
> "How much the feeling that he was of inferior intelligence, mathematically, to Fermat, contributed to his decision to give up mathematics it is impossible to say, ...;"

138

"If one looks on the first calculations of a probability by Cardan and by Galileo as unimportant and says that the real begetter of the calculus of probabilities is he who first put it on a sound footing, then we should pass over not only Cardan and Galileo but also Pascal and Fermat. For although Fermat, if provoked, could have done as much and more than his successors, yet the fact remains that his contribution was in effect the extension of the idea of the exhaustive enumeration of the fundamental probability set, which had already been given by Galileo. The scientist who first put forward in a systematic way the new propositions evoked by the problems set to Pascal and Fermat, who gave the rules and who first made definitive the idea of mathematical expectation was Christianus Huygens."

In the last of these quotations David omits any mention of Pascal's contribution, and then attributes (as do many writers) the defining of the notion of expectation to Huygens, though not only was it used first by Pascal, but Huygens knew this fact before his book *De ratiociniis in aleae ludo* (1657) was printed, as we shall see. There is no doubt that David's opinion was much influenced by her unfamiliarity with the relevant tract appended to Pascal's book *Traité du triangle arithmétique* (1665) in which he gave his complete solution to the Problem of Points, making use of expectations to obtain it.

Rényi (1972) put in a word for Pascal:

"According to David, the solution of de Méré's (and herewith the theoretical foundation of the calculus of probabilities) is due to Fermat; her reasoning, however, is not completely convincing. The witty solution of the problem of rightful distribution (the Problem of Points) comes certainly from Pascal, and this, a recursive procedure making superfluous the actual counting of the individual cases, is in itself a significant contribution to the foundation of probability theory;"

and Hacking in *The Emergence of Probability* (1975) refers to Pascal as "the founder of modern probability theory" and writes that "Not until the correspondence between Fermat and Pascal do we find expectation well understood".

It is clear from these references that a fresh examination of the evidence is now needed, and the present paper is intended to provide it. Accounts of the Problem of Points other than those already mentioned are given by Maistrov (1974), Sheynin (1977), and in the introduction to Volume XIV of Huygens's Collected Works (Huygens, 1920).

2   Pascal–Fermat correspondence

The Problem of Points, also known as the "division problem", involves determining how the total stake should be equitably divided when a game is terminated prematurely. Suppose two players A and B stake equal money on being the first to win n points in a game in which the winner of each point is decided by the toss of a fair coin, heads for A and tails for B. If such a game is interrupted when A still lacks a points and B lacks b, how should the total stake be divided between them?

Particular cases of the problem have been noted in Italian mathematical manuscripts as early as 1380 (Ore, 1960), and the Renaissance mathematicians Paciolo, Tartaglia and Peverone all made unsuccessful efforts at solutions, Tartaglia concluding that "the resolution of such a question is judicial rather than mathematical, so that in whatever way the division is made there will be cause for litigation".

The problem was brought to Pascal's attention by Antoine Gombaud, chevalier de Méré, and led to a correspondence with Fermat in the late summer of 1654. Unfortunately not all the letters have survived, but it is possible to reconstruct the sequence of events from the remainder, which are available in English translation (David, 1962, Appendix 4; Smith, 1929). The originals are printed in *Oeuvres de Fermat* (Tannery & Henry, 1894) and in *Pascal: Oeuvres complètes* (Mesnard, 1970).

On 28th July Pascal received a letter from Fermat which evidently contained solutions to two gaming problems, one of which was the Problem of Points. This letter must have been preceded by one from Pascal, also missing, in which he set out the problems and possibly also his solutions to them. On 29th July he replied at length on the Problem of Points, agreeing with Fermat's solution (through a most unfortunate error of translation (Smith, 1929, p. 547; David, 1962, p. 231) it has been represented that Pascal and Fermat originally disagreed over the solution). Pascal noted that de Méré had been unable to obtain a solution himself. He writes

"Your method is sound and is one which first came to my mind in this research; but because the labour of the combinations is excessive, I have found a short cut and indeed another method which is much quicker and neater, which I would like to tell you here in a few words."

The method of combinations, to which he here refers, and which he and Fermat had thus apparently thought of independently, is as follows. (We do not know whether the earlier letters treated a specific example or gave a general solution, but though the former seems more likely we shall reconstruct the latter.) If A wants a points and B wants b then the game must be over in at most $(a + b - 1)$ further tosses, which may occur in 2^{a+b-1} equiprobable ways. Each of these ways may be seen, on examination, to be a win either for A or for B, and on counting them the probability (as we should now say) of A winning is found, and hence the equitable division of the stakes given that the game is not in fact continued.

This solution amounts to no more than an exhaustive enumeration of all the possibilities, and as such introduces no principle not known to Cardano and Galileo, but it does have two interesting features. First it implicitly assumes that the total length of the game as originally intended is irrelevant: all that matters are the possibilities that remain. This important principle was, however, also implicit in the erroneous solutions of earlier writers. Secondly although the enumeration of the 2^{a+b-1} equiprobable endings is trivially straightforward, it is to be noted that had the game continued it might not have lasted for as many as $(a + b - 1)$ further tosses, so that someone with less of a feel for probability than Pascal and Fermat might object that their method of combinations includes, in its enumeration, tosses that would under no circumstance have been made. Roberval was to make just this point when he discussed the method with Pascal. To us, of course, there is no paradox, and we may instead observe that Pascal and Fermat have hit upon the notion of "embedding" one probability problem in another.

However, it was not this paradox which inspired Pascal to seek an alternative solution at the outset, but the tedium of enumerating all the possibilities. He must have had it ready before he received Fermat's letter on 28th July, because on the following day he described his alternative (and much more) at length in his reply. This solution is of the greatest importance to the development of probability theory because in it Pascal introduces the notion of expectation. In modern terms his solution involves analysing the tree of possible games and, working backwards from the tips, using recursively the idea that, if expectations of gain of X and Y units are equally probable, the expectation of gain is $\frac{1}{2}(X + Y)$ units. Furthermore, the stakes are to be divided according to the expectation of gain, that is, the value of a gamble is equal to its expectation.

Pascal did not generalize his method fully in the correspondence, but his letter does solve two partial generalizations. The first is the case in which A wants but one point ($a = 1$) whilst B wants many (b). We should now argue that in order for B to win the game he must win each of the next b tosses, an event with probability $1/2^b$, which is therefore the proportion of the total stake he should receive in compensation for giving up the unequal struggle! Pascal saw this solution inductively, as can we by drawing an event-tree.

The second partial generalization solved by Pascal is the case in which one player is just one point ahead of the other, so that if B wants b points, A wants $(b - 1)$. However, he was unable to reach the solution through his method of expectations (an attempt to draw the corresponding event-tree will show why: it is as complicated as in the completely general case), and proceeded instead to develop the method of combinations. In order to appreciate the important advance he here makes, it is necessary to emphasize a remarkable feature of the method of combinations applied to the general Problem of Points when there are just two players: *the order of occurrence of the heads and tails does not matter*, but only the total numbers of each, notwithstanding the fact that the game is not of determinate length. The reason is that, in the imaginary game of $(a + b - 1)$ tosses, even if A only scores the a heads he needs to win, and no more, B, playing on to the bitter end, would only score $(b - 1)$ tails, one short. And if A scores more than a heads, B is even further away from winning. In other words if the imaginary game contains $0, 1, 2, \ldots, (a - 1)$ heads *in any order* B will have won the game, whilst if it contains $a, (a + 1), (a + 2), \ldots, (a + b - 1)$ heads

A will have won the game. Playing on so as to complete $(a + b - 1)$ tosses makes no difference. Representing the game on a lattice diagram (see Fig. 22 and 23, page 73) makes this clear.

Understanding this, Pascal is then able to make use of the fact that in $(a + b - 1)$ tosses 0, 1, 2, ... heads will occur in $^{a+b-1}C_0$, $^{a+b-1}C_1$, $^{a+b-1}C_2$, ... ways, each equally probable, the numbers being given by the row of the arithmetical triangle corresponding to the expansion of $(1 + 1)^{a+b-1}$. With his knowledge of the properties of these binomial coefficients he is able to show that if B wants b points and A wants $(b - 1)$ the correct division of the stake is

$$A:B = 2^{2b-2} + {}^{2b-2}C_{b-1}:2^{2b-2} - {}^{2b-2}C_{b-1}.$$

As usual, Pascal has actually worked with a particular example, but the general method is clear. Taking $b = 5$, he has found the division to be $2^8 + 70:2^8 - 70$.

Apart from his skill at manipulating binomial coefficients, Pascal has here demonstrated his understanding of the fact that, with two players, the 2^{a+b-1} possible imaginary games can be aggregated, through the medium of binomial coefficients, into the ones that A wins and the ones that B wins. It is thus only necessary to compute

$$^{a+b-1}C_0 + {}^{a+b-1}C_1 + \ldots + {}^{a+b-1}C_{a-1}:$$
$$^{a+b-1}C_a + \ldots + {}^{a+b-1}C_{a+b-1}$$

as the division ratio $A:B$, and this is the solution to the general problem. However, Pascal did not give this explicitly in his letters to Fermat (though he was to prove it by mathematical induction in the *Traité du triangle arithmétique*), presumably because he did not know (and we do not know) of any simple expression for the partial sum of binomial coefficients, and he therefore limited himself to solving the almost-symmetrical case $a = b - 1$ described above in which the partial sum may be simply expressed.

This solution of the general case in terms of binomial coefficients has often been attributed to Fermat, but there is no evidence that Fermat, in the method of combinations, ever did more than enumerate all the 2^{a+b-1} possibilities (which Pascal found excessively laborious). Indeed, in a later letter, dated 24th August, Pascal describes Fermat's procedure in detail for the case $a = 2$ and

$b = 3$ (the case given in Fig. 22 and 23, page 73) and enumerates all the $2^{a+b-1} = 16$ possibilities, not even grouping them according to the numbers of heads in each.

Pascal continues his letter by giving tables in which he has computed the division of the stakes for all the possible situations (a, b) from $(0, 1)$ to $(5, 6)$, the second table being the cumulative version of the first. These tables may have been constructed equally well by either of his methods. He refers to the fact "that the value of the first game is equal to the second, which is easily shown by combinations," and this he does, by manipulating binomial coefficients, in Proposition IV of the *Usage du triangle arithmétique pour déterminer les partis...*, a tract appended to the *Traité du triangle arithmétique*.

Fermat's reply to Pascal's letter dated 29th July is not extant (and may never have existed) but there was certainly a letter to which Pascal's letter of 24th August was itself a reply because in its closing paragraphs it refers to Fermat having propounded the Problem of Points for several players. Indeed, it is likely that a letter of Pascal's on this topic, dated about 20th August, is also missing.

In order to understand the remaining letters fully, it is necessary to follow precisely what the correspondents meant by the "combinatorial method" which, as we have seen, Pascal and Fermat had independently devised as their solution to the problem. As the correspondence progresses Pascal refers to it as Fermat's method, not because he did not also think of it himself (as he mentions in letters of 29th July and 24th August), but in contrast to *his* method of expectations.

Pascal and Fermat took different views as to how the combinatorial method was to be extended to three players, and this has led commentators to assert that Pascal made a mistake. The truth is more complex: Pascal had extended *what he thought Fermat meant by the combinatorial method* to a case involving three players, and had obtained an answer which he knew was wrong since it did not correspond to the solution obtained by his method of expectations. In his letter of 24th August he not only showed Fermat his detailed workings, but explained to him where the fault in the argument lay. He had realized that, in the hypothetical game in which the real game is embedded in the method of combinations, the order of heads and tails is only irrelevant with two players, and not, in general, with three or more.

Pascal had supposed that ignoring the order in which heads and tails occurred was an intrinsic feature of the combinatorial method—after all (we may note), that is what is meant by a combination as opposed to a permutation. In his letter he works carefully through his original example for two players, listing all the permutations for each combination, but then assigning each to A or B according to the numbers of heads and tails, as is justifiable with only two players. Then he applies the identical argument to three players, and immediately notes that some *combinations* are favourable to two players, but that to count each corresponding permutation as a win for both would be to make "a gross erros". Therefore, he supposes, the method of combinations must proceed by dividing each such permutation equally between the two players, counting it as half a win for each. He writes to Fermat "That seems to me to be the way one must solve the problem of points following your combinatorial method, unless you proceed in some other way which I do not know. But, if I am not mistaken, this solution is unfair". He then explains the difficulty: "In the actual conditions of the game with three players, only one man can win, for play ceases as soon as one man has won. But in the hypothetical conditions, two men can get the number of games they need."

Thus Pascal is pointing out that with three players one cannot solve the problem by only counting the *numbers* of points won by each player in the hypothetical game; in other words, one cannot in general solve the Problem of Points for three players by what he conceives to be the "combinatorial method" (he allows that in particular cases it might work).

In his reply on 25th September Fermat makes the obvious rejoinder, that his method entails assigning each permutation of each combination to one of the players, and that it will then work for three players. The misunderstanding is happily at an end. Perhaps the root of the problem was the ambiguity in the word *combinaison*, and when in the *Traité du triangle arithmétique* Pascal introduces it he says "The word *combinaison* has been used in many different senses, so that in order to avoid ambiguity I find it necessary to state exactly what I intend by it," and he goes on to specify that in his

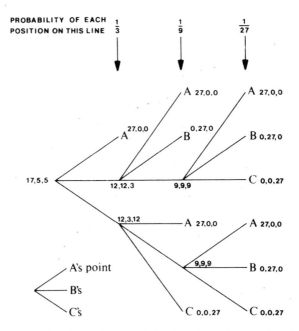

Fig. A1 Event-tree for the Problem of Points for three players when *A* needs one point to win and *B* and *C* both need two. A stake of 27 is available to the winner (this number is chosen for ease of calculation). The tips of tree are labelled with winners, and associated number triplets show the winnings of *A*, *B* and *C* respectively, whilst at nodes they show expected winnings at corresponding stages of the game calculated by Pascal's argument. Probabilities show how Fermat calculated directly the probability of *A* winning to be $1 \times \frac{1}{3} + 2 \times \frac{1}{9} + 2 \times \frac{1}{27} = \frac{17}{27}$

usage *order will not be taken into account.*

Also in his letter of 25th September Fermat gives, for the benefit of Roberval, the direct probability solution for the game of three players in which one man lacks one point and the others two each, simply working his way *forwards* along the event-tree assigning probabilities, as we would now expect a student to do. Notwithstanding the fact that neither Pascal, with his backwards expectation argument, nor Fermat, with this forward probability argument, ever drew the event-tree as far as we know, Fig. A1 may clarify the two methods. The first drawing of an event-tree I have seen is in a manuscript of Huygens dated 1676; see Huygens (1920). Pascal, in his reply of 27th October, was pleased with Fermat's letter which restored their "mutual understanding", and especially with the direct probability solution which, he noted, was entirely Fermat's and had nothing in common with his.

3 Conclusion

Thus does the famous correspondence end, on a happy note. What has been achieved? Not one but three correct arguments have been advanced to solve the Problem of Points for any number of players. The first method of solution is to find the maximum number of tosses which will be needed to determine the winner, then list all the possible sequences of heads and tails in this number of tosses, and determine by inspection who is the winner in each case. This way of looking at the problem may be attributed jointly to Fermat and to Pascal, and Pascal's belief that Fermat had intended it to be applied to three players just as it had been applied to two (by ignoring the order of occurrence of the tosses) may be dismissed as an unimportant misunderstanding. The second method, due wholly to Pascal, is to take expectations, working backwards, and the third method, due to Fermat but involving no new principle, is to enumerate the possibilities working forwards. In addition, for the case of two players only, Pascal has understood that the *combinations* favourable to one player are enumerated by the partial sum of the appropriate row of the arithmetical triangle of binomial coefficients. This is the solution which, on no evidence whatever, is often attributed to Fermat. Although it is only implicit in Pascal's letter (of 29th July) it is not only given explicitly in the tract, already mentioned, appended to the *Traité du triangle arithmétique*, which Fermat had received by 29th August, but Pascal there *proves* the result for two players by mathematical induction applied to his method of expectation. Specifically, assuming the expectation for each player after a particular toss to be given by this result, he computes the expectations before the toss by taking the means of the expectations for the two cases A wins the toss and B wins it, and shows that they conform to the result; but it is certainly true when each player needs a single point ($a + b = 2$) and therefore it is true for $a + b = 3, 4, 5, \ldots$.

In these few months in the summer of 1654 the theory of probability was taken beyond the simple enumeration of equally-probable events, the so-called fundamental probability set. A famous unsolved problem was solved, not merely by the discovery of the appropriate fundamental probability set and by its correct enumeration, but by Pascal's introduction of the expectation argument, his application of mathematical induction, and his implicit use of the binomial distribution for equal chances.

Pascal's binomial solution for two players was given as a formula by Pierre Montmort in the first edition (1708) of his *Essay d'Analyse sur les Jeux de Hazard*. The similar formula for the case of players of unequal skill (or, equivalently, a biased coin) was communicated to Montmort by John Bernoulli in a letter dated 17th March 1710, and given in Montmort's second edition (1713), where an alternative formula is also given: that which arises when the probabilities of the actual possible games are computed, rather than those of the hypothetical games leading to the binomial solution (see Fig. 22). Some of the tendency for the binomial solution to be attributed to Fermat may be traced to Montmort (1713), for though he correctly draws a distinction between the method of combinations and Pascal's method of expectations, he develops the former using binomial coefficients and attributes the resulting formula to Fermat. De Moivre, in *De Mensura sortis* (1712), was the first to *publish* the binomial solution for two players of unequal skill, and he goes on to give the similar solution for three players, noting carefully that some of the terms of the trinomial distribution have to be split up because the winner is determined by the order in which the results were obtained. Montmort had done as much in his letter to Nicholas Bernoulli dated 8th June 1712, before he received De Moivre's book at the beginning of August, though Bernoulli, in his reply, pointed out that Montmort has missed a term (Montmort, 1713).

Huygens's famous *De ratiociniis in aleae ludo* (1657) included the Problem of Points for any number of players of equal skill and gave Pascal's expectation method as the means of solution. Huygens had learnt of de Méré's problems during a visit to Paris in 1655, in the year following the Pascal–Fermat correspondence, and wrote in the foreword to his book (David, 1962):

"It should be said, also, that for some time some of the best mathematicians of France have occupied themselves with this kind of calculus so that no one should attribute to me the honour of the first invention. This does not belong to me. But these savants, although they put each other to the test by proposing to each other many questions difficult to solve, have hidden their methods. I have had therefore to examine and to go deeply for myself into this matter by beginning with the elements, and it is impossible for me for this reason to affirm that I have even started from the same principle. But finally I have found that my answers in many cases do not differ from theirs."

It is *just* possible to believe that no inkling of Pascal's method of expectations was communicated to Huygens along with the Problem of Points, but difficult. In any event, through correspondence with Carcavi in the autumn of 1656 (Huygens, 1888) he learnt that his method was the same as Pascal's, as a result of which he added Proposition IX to his book, giving Pascal's method applied to the Problem of Points succinctly, so that before his book went to press he knew of Pascal's priority in the matter of expectations. (I do not intend any criticism of Huygens by these remarks; the obligation to quote others was not then a firm part of mathematical writing.)

When James Bernoulli opened his *Ars conjectandi* (1713; Bernoulli died in 1705) with a commentary on Huygens's treatise he was naturally led to some elaboration of the problem, but he contributes nothing not known to Pascal, of whose *Traité du triangle arithmétique* he was oblivious until just before he died (Haussner, 1899). However, in an adjoined tract *Lettre a un Amy, sur les Parties du Jeu de Paume* Bernoulli applies probability theory to the game of tennis (which is only a glorified Problem of Points!) in the course of which he introduces players of unequal skill; since he worked on this in the winter of 1685–6 (Bernoulli, 1975) it seems probable that the formula communicated to Montmort by John Bernoulli in 1710 originated with the latter's brother.

References to Appendix 1

Bernoulli, J. (1713). *Ars conjectandi*. Basilea: Thurnisius. Reprinted (1968) Bruxelles: Culture et Civilisation, and Bernoulli (1975).

Bernoulli, J. (1975). *Die Werke von Jakob Bernoulli*, III. Basel: Birkhäuser.

David, F. N. (1962), *Games, Gods and Gambling*. London: Griffin.

De Moivre, A. (1712). De mensura sortis. *Phil. Trans. R. Soc. Lond.* **27**, 213–64.

Hacking, I. (1975). *The Emergence of Probability*. Cambridge University Press.

Haussner, R. (1899). *Wahrscheinlichkeitsrechnung (Ars conjectandi von Jakob Bernoulli, 1713). Ostwald's Klassiker* 107. Leipzig: Engelmann.

Huygens, C. (1657). *De ratiociniis in aleae ludo*. In *Excercitationum mathematicarum* by F. van Schooten. Lugduno Batava: *ex* Johannis Elsevirii. Reprinted with commentary as Part I of Bernoulli (1713) and in Dutch and French in Huygens (1920).

Huygens, C. (1888). *Oeuvres complètes de Christiaan Huygens*, I. La Haye: Nijhoff.

Huygens, C. (1920). *Oeuvres complètes de Christiaan Huygens*, XIV. La Haye: Nijhoff.

Mahoney, M. S. (1973). *The Mathematical Career of Pierre de Fermat*. Princeton University Press.

Maistrov, L. E. (1974). *Probability Theory: A Historical Sketch*. New York: Academic Press.

Mesnard, J. (Ed.) (1970). *B. Pascal: Oeuvres complètes*, II. Bruges: Desclée De Brouwer.

Montmort, P. de (1708). *Essay d'analyse sur les jeux de hazard*. Paris: Quillau.

Montmort, P. de (1713). *Essay d'analyse sur les jeux de hazard*, 2nd edn. Paris: Quillau. Reprinted (1980), New York, Chelsea.

Ore, O. (1953), *Cardano, the Gambling Scholar*. Princeton University Press. Reprinted (1953), New York: Dover.

Ore, O. (1960). Pascal and the invention of probability theory. *Am. Math. Mon.* **67**, 409–19.

Pascal, B. (1665). *Traité du triangle arithmétique*. Paris: Desprez. Reprinted in Mesnard (1970).

Rényi, A. (1972). *Letters on Probability*. Detroit: Wayne State University Press.

Sheynin, O. B. (1977). Early history of the theory of probability. *Arch. Hist. Exact Sci.* **17**, 201–59.

Smith, D. E. (1929). *A Source Book in Mathematics*. New York: McGraw-Hill.

Tannery, P. and Henry, C. (Eds) (1894). *Oeuvres de Fermat*. II. Paris: Gauthier-Villars.

Todhunter, I. (1865). *A History of the Mathematical Theory of Probability*. Cambridge: Macmillan. Reprinted (1965), New York: Chelsea.

Appendix 2

Pascal's Problem: The "Gambler's Ruin"

Reprinted from *Int. Statist. Rev.* **51**, 73–9 (1983) by permission of the International Statistical Institute

1 Introduction

Some time in 1656, two years after his famous correspondence with Pierre de Fermat on the Problem of Points, Blaise Pascal posed Fermat another problem in probability, which he considered so difficult that he doubted whether Fermat would be able to solve it; in his view it was more difficult than all the others they had discussed (Carcavi, 1656).

Let two men play with three dice, the first player scoring a point whenever 11 is thrown, and the second whenever 14 is thrown. But instead of the points accumulating in the ordinary way, let a point be added to a player's score only if his opponent's score is nil, but otherwise let it instead be subtracted from his opponent's score. It is as if opposing points form "pairs", and annihilate each other, so that the trailing player always has zero points. The winner is the first to reach twelve points; what are the relative chances of each player winning?

This is Carcavi's description (freely rendered), and there is no reason to suppose it much different from what Pascal had proposed. When Huygens heard it he immediately thought of the problem in terms of the players' points accumulating in the ordinary way, but the winner being he who first *led* by twelve points (Huygens, 1656), and when he put it in his *De ratiociniis in aleae ludo* (Huygens, 1657) he rendered it in terms of each player starting with twelve points, a win for a player then involving the transfer of a point from his opponent to himself, the overall winner being he who bankrupts the

151

other of points. All three forms of this problem of Pascal's are, of course, equivalent, but it is in the last form that the problem has come down to us, as the "Gambler's Ruin" problem.

In *De ratiociniis* Huygens gave the problem's solution (agreeing with Pascal's, as we shall see), the reader to supply the demonstration, and the problem became even better known through James Bernoulli's annotated edition of *De ratiociniis* which forms Part I of his *Ars conjectandi* (Bernoulli, 1713). Bernoulli experienced some difficulty in his attempts to justify the solution, thus apparently vindicating Pascal's opinion of the difficulty of the problem.

That the name of Pascal is not normally associated with the Gambler's Ruin problem is easily explained: Huygens did not mention his source (and Bernoulli probably did not know it), which was only revealed with the publication of Huygens' correspondence in 1888. Since this was after Todhunter wrote his *History of the Mathematical Theory of Probability* (Todhunter, 1865) he does not record the origin either, and his book has been the major secondary source of information about the early history of probability ever since. I do not know if the problem has any history prior to Pascal; I have not seen it mentioned any earlier, though it seems a perfectly natural problem for gamblers seeking a new game to play. It has a special importance in the development of probability theory because it led mathematicians at the beginning of the eighteenth century to investigate the duration of play, which turned out to be a very fruitful topic. "The last of the five problems which Huygens left to be solved", wrote Todhunter, "is the most remarkable of all. It is the first example on the *Duration of Play*, a subject which afterwards exercised the highest powers of De Moivre, Lagrange, and Laplace"; he considers their contributions in his book, but for more recent accounts see Thatcher (1957) (who was, however, also unaware that the problem originated with Pascal) and Kohli (1967). The purpose of the present account, apart from drawing attention to the origin of the problem, is to speculate on how Pascal and Fermat solved it, for they left no indication.

We may make the preliminary observation that none of the solvers had any hesitation in giving the answer to the first part of the problem, namely, that the odds of scoring 11 versus 14 with three dice are 9:5.

2 Pascal's solution

To Pascal's chagrin, Fermat returned a correct solution via Carcavi immediately: the odds, he said, were between 1156:1 and 1157:1, and Pascal, fearing that he might be thought to have posed Fermat a problem which he had not solved himself, forthwith revealed the exact solution to Carcavi:

150, 094, 635, 296, 999, 122:129, 746, 337, 890, 625.

We know nothing about the methods of solution; only that Pascal and Fermat "seem to have used different methods" (Carcavi, 1656). Evidence for what these were is therefore limited to the correspondence on the Problem of Points two years earlier, where we can study the different styles of the two men (Edwards, 1982); in particular we learn that Fermat liked to enumerate all the possibilities in a chance process, and was also happy to advance a direct probability argument using implicitly both the multiplication and addition rules, whilst Pascal had invented a recursive argument applied to his concept of expectation.

The Problem of Points is as follows. Let two players A and B stake equal money on being the first to win n points in a game in which the winner of each point is decided by the toss of a fair coin, heads for A and tails for B. If such a game is interrupted when A still lacks a points and B lacks b, how should the total stake be divided between them?

Pascal's method, invented in 1654 (and sent to Fermat) but not published until after his death (Pascal, 1665), involves the notion of expectation. Let $E(a, b)$ be the expectation, or expected winnings, of player A when the game stands at (a, b). Then Pascal asserted that

$$E(a, b) = \tfrac{1}{2}E(a - 1, b) + \tfrac{1}{2}E(a, b - 1) \tag{1}$$

and used this to prove his solution for $E(a, b)$ which he already knew from combinatorial arguments involving the binomial coefficients (Edwards, 1982).

Pascal's new problem is, of course, a modified Problem of Points, where the absolute progress of the players is replaced by relative progress (and unequal probabilities of winning each point are thrown in for good measure), and I conjecture that Pascal invented this generalization in order to display his method of expectations in a problem which would not (he wrongly supposed) be amenable to more direct solution: it would be a good challenge to his friend Fermat, who would find himself trying to enumerate sequences of indefinite length!

Corresponding to equation (1) Pascal would have

$$E(a, b) = pE(a + 1, b) + qE(a, b + 1), \tag{2}$$

where A has scored a points overall (each with probability p) and B has scored b (each with probability q), and, in accordance with Carcavi's description of the problem

$$E(a, b) = \begin{cases} E(a - b, 0) & (a \geqslant b), \\ E(0, b - a) & (a \leqslant b), \end{cases} \quad E(12, 0) = 1, \quad E(0, 12) = 0, \tag{3}$$

where the total stake is taken to be 1 for convenience. It is immediately obvious that charting the game on the scale $(12, 0)$ to $(0, 12)$ is a tiresome version of charting it on the scale (0) to (24), where this number refers to the number of points B (say) needs to win straight off in order to win the game; therefore let $E(0, 12) = E_0 = 0$, $E(0, 11) = E_1$, $E(0, 10) = E_2, \ldots, E(0, 0) = E_{12}$, $\ldots, E(12, 0) = E_{24} = 1$. Pascal would then have

$$E_0 = 0, \quad E_1 = pE_2 + qE_0, \quad E_2 = pE_3 + qE_1, \quad \ldots, \quad E_{24} = 1, \tag{4}$$

25 equations in 25 unknowns, with a solution required for E_{12}.

Equation (4), if cast into modern form, is simply a second-order homogeneous linear difference equation complete with two boundary conditions, and may be solved by the usual methods (Feller, 1957). But these methods were not available to Pascal and his contemporaries; at a later stage Hudde (1665), Bernoulli (1684) and Montmort (1708) all solved (4) by direct substitution, eliminating the unwanted unknowns one by one, but I suggest that

Pascal employed a trick which he had already used to advantage to sum the powers of the integers in a tract printed with the *Traité du triangle arithmétique* (Pascal, 1665; see Edwards, 1983), a trick in which successive equations are thrown into such a form that when they are added together many of the terms cancel out. Thus if he replaced E_1 in the second equation by $pE_1 + qE_1$ and rearranged it, and so on, he would have

$$pE_2 - pE_1 = qE_1 - qE_0, \quad pE_3 - pE_2 = qE_2 - qE_1,$$

$$pE_4 - pE_3 = qE_3 - qE_2, \quad \ldots, \quad pE_m - pE_{m-1} = qE_{m-1} - qE_{m-2}, \tag{5}$$

which on being summed give simply

$$pE_m - pE_1 = qE_{m-1}, \tag{6}$$

since $E_0 = 0$. This equation being first order, successive substitution may be accomplished algebraically:

$$E_m = \frac{q}{p} E_{m-1} + E_1,$$

$$= \left(\left(\frac{q}{p} \right)^{m-1} + \left(\frac{q}{p} \right)^{m-2} + \ldots + \left(\frac{q}{p} \right)^2 + \frac{q}{p} + 1 \right) E_1$$

$$= \frac{1 - \left(\dfrac{q}{p} \right)^m}{1 - \dfrac{q}{p}} E_1, \tag{7}$$

since the series to be summed is geometric.

All that is now needed is to put $m = 24$ in (7), thus solving for E_1, and then to put $m = 12$ to find

$$E_{12} = \left\{ 1 - \left(\frac{q}{p} \right)^{12} \right\} \bigg/ \left\{ 1 - \left(\frac{q}{p} \right)^{24} \right\}. \tag{8}$$

Similar reasoning leads to the expectation for player A, and the full answer is then, after simplification, that the odds are

$$p^{12}:q^{12}. \tag{9}$$

Putting $p = 27$ and $q = 15$ (Pascal evidently failed to cancel out the common factor 3) leads to Pascal's exact answer quoted above.

Thus I suggest that Pascal moved from using an induction based on expectations to prove his earlier solution to the Problem of Points (1) to using the comparable equation (2) for the Gambler's Ruin problem to obtain a hitherto unknown solution by an application of a technique (summing equations) which he is known to have used previously. Not only was he technically capable of such a development, but this suggestion explains why he doubted Fermat would be able to find the solution: if he had himself used a simpler method, that doubt would not have been well founded. Dazzled by his own solution he failed to notice a simpler argument; but Fermat, concerned with solving the problem and not with exemplifying a method, found one immediately the problem was put to him.

3   Fermat's solution

There is no reason to suppose that Fermat would have proceeded by expectations at all; that was Pascal's method, and did not become standard until Huygens codified it in *De ratiociniis*. Much more characteristic would have been an attempt to list all the possible games, as follows.

First, as with the Problem of Points, Fermat will have asked himself the maximum number of throws required to complete the game, and will have been alarmed (we may suppose) to discover that the game could continue indefinitely. Realizing that the chance of a very long game was small, Fermat will then have decided to analyse the games starting with the shortest. If A takes 12 points straight off (probability p^{12}) he wins, as does B (probability q^{12}); in this, the shortest possible game (12 throws), the odds are therefore $p^{12}:q^{12}$. The next shortest game has 14 throws, 13 points for one player and 1 for the other, but when Fermat enumerated the permutations he will have seen that two of the fourteen lead, in each case, to a win at the twelfth throw, and therefore that the odds are $12p^{13}q:12pq^{13}$, or

$p^{12}:q^{12}$ again. The next shortest game has 16 throws, each permutation leading to a win for A having probability $p^{14}q^2$ and each one leading to a win for B having probability p^2q^{14}, and, however many permutations there are in each case, symmetry requires the same number amongst the wins for A as amongst the wins for B. Once again therefore, the odds are $p^{12}:q^{12}$. The argument is obviously quite general: games of any length command odds of $p^{12}:q^{12}$, and therefore (whatever the probability distribution of lengths, we should now say) the odds for the game as a whole will be $p^{12}:q^{12}$. Probably using logarithms, Fermat computed this for $p = 9$ and $q = 5$ and returned the correct answer to Pascal. Indeed, if he possessed at least five-figure logarithms of four-figure numbers the result "between 1156:1 and 1157:1" will have been the most natural format.

I suggest that this argument was well within Fermat's capacity, accords with what we know about his thinking in games of chance, and explains both the speed and form of his answer. We may note that it is neither as general as Pascal's suggested method (which includes the solution where the players are set unequal numbers of points by which to lead in order to win), nor does it incidentally demonstrate that the probability of the game continuing for ever is zero (which Pascal's does because the sum of the probabilities of winning is unity); but it answered Pascal's challenge, which was all that mattered.

4   Huygens' solution

As we have noted, Huygens did not attach any demonstration or explanation when he gave the problem and its solution (Pascal's solution, with the common factor of 3^{12} removed) in *De ratiociniis*. Nor had he enlightened Carcavi about his method when he wrote to him with the solution (Huygens, 1656), except to mention that he had used expectations "and algebra". But in August 1676 he described a method on a sheet headed "The last problem contained in *De ratiociniis*; proposed some time ago by Pascal" (Huygens, 1676), and there is no reason to suppose it any different from the one he had used 20 years previously. First he considers the case in which the winner is he who first leads by two points (see §1 above for this form of the game), and solves the corresponding equations (4) algebraically for E_2, in our terminology. It is interesting to note the use of an "event-tree" to describe the possible events, perhaps the first time such a diagram had been used.

Huygens next tackles the case where a lead of four points is required to win, which he cleverly solves by considering only every other possible state of the game, namely the points $(4, 0)$, $(2, 0)$, $(0, 0)$, $(0, 2)$ and $(0, 4)$, and noting that the event-tree then looks like the first one, save that all the chances are squared. The justification for leaving out the intermediate points is evidently that to pass from $(0, 0)$ to $(0, 4)$, for example, it is necessary to pass through $(0, 2)$. By a sort of convolution of the two-points problem into itself, therefore, the relative chances will be $(p^2)^2 : (q^2)^2 = p^4 : q^4$. The argument is by no means obvious, and since Huygens offers no explanation beyond a diagram we cannot tell exactly what form it took in his mind.

A further application of the argument takes Huygens to $p^8 : q^8$ for the case where a lead of eight points is required to win, and by a variant of the argument he obtains $p^3 : q^3$ and thence $p^6 : q^6$. He then says that the solution $p^5 : q^5$ could be obtained in a similar way, though it would take longer, and concludes that the general solution, for the case where a lead of n points is needed to win, must be $p^n : q^n$. This argument reappears in James Bernoulli's *Ars conjectandi* (Bernoulli, 1713), where the author says that if $P : Q$ is the solution when each player starts with n points (he considers the third form of the game) then the solution if each starts with $2n$ points will be $P^2 : Q^2$; hence we may proceed from the known solution $p^3 : q^3$ ($n = 3$) to $p^6 : q^6$ and thence to the solution of Pascal's case $p^{12} : q^{12}$. But Bernoulli gives no reason why this induction should be valid.

5   Other solutions

We noted briefly above that Hudde (1665) solved equations (4) by direct substitution, doing so algebraically for the cases where the players start with 1, 2 and 3 points each. This solution, identified as Hudde's by the handwriting, was found amongst Huygens' papers. The solution for the 3-point case is direct, in contrast to Huygens' solution which we mentioned as involving a convolution. Hudde's solution also reappears in *Ars conjectandi*; indeed, it is the first one that Bernoulli gives, adding the general rule $p^n:q^n$ by a natural extension. "But", he goes on, "this could actually be inferred without calculation at all", and he proceeds to repeat an unsatisfactory explanation from 1684 (Bernoulli, 1684) that confuses the *probability of winning* with the *probability of winning at the nth toss*, which is of course p^n for one player and q^n for the other. Kohli (1967) attempts to rescue Bernoulli by supposing that he had in mind the kind of argument we attributed to Fermat in §3 above, but I think that is too charitable an interpretation. Bernoulli was struggling, and he knew it, for he continues "But if, however, anyone considers this reasoning unsatisfactory, and is also not persuaded by the induction [that is, the 'natural extension' of Hudde's argument], he may proceed by the following short cut . . ., passing immediately [from the case $n = 3$] to the case $n = 6$ and thence to $n = 12$, omitting all the intermediate cases" (see §4 above).

Finally, Bernoulli quotes the general solution for the game in which A starts with m points and B with n, which is

$$p^n(p^m - q^m):q^m(p^n - q^n), \tag{10}$$

but any hope the reader might have that Bernoulli is then going to subsume his earlier arguments into a general demonstration is quickly dashed: "The demonstration of this is left for the reader to expound, even though he earnestly requests the rather troublesome calculation. But we will pass on to another part of our theme without further delay". These are the closing words of Part I of *Ars conjectandi*, and leave a distinct feeling that Bernoulli was not too happy with what he had written. We must, however, remember that this work was not published during its author's lifetime, and he might have hoped to have improved on his treatment of Pascal's problem before publication.

The general solution (10) is readily obtained by the method we described in §2 and suggested Pascal might have used to solve his special case, but it is not clear who first found it. In his letter to Montmort dated 26 February 1711 (in Montmort, 1713) Nicholas Bernoulli (who prepared *Ars conjectandi* for press) mentioned the solution in connection with investigating the duration of play, but added that he had derived it previously by another method. The first actual publication of (10) was by De Moivre (1712), whose elegant proof is given by Thatcher (1957).

References to Appendix 2

Bernoulli, James (1684). *Die Werke von Jakob Bernoulli*, III. Basel 1975: Birkhäuser.

Bernoulli, James (1713). *Ars conjectandi*. Basilea. Facsimile reprint (1968), Bruxelles: Culture et Civilisation. Reprinted (1975) in Bernoulli's *Werke* (see Bernoulli, 1684), pp. 107–286.

Carcavi, P. (1656). In Huygens (1656).

De Moivre, A. (1712). De mensura sortis. *Phil. Trans.* **27**, 213–64.

Edwards, A. W. F. (1982). Pascal and the Problem of Points. *Int. Statist. Rev.* **50**, 259–66.*

Edwards, A. W. F. (1983). *Pascal's Arithmetical Triangle.* To appear.†

Feller, W. (1957). *An Introduction to Probability Theory and its Applications*, 1, 2nd edn. New York: Wiley.

Hudde, J. (1665). *Oeuvres complètes de Christiaan Huygens*, V: *Correspondance 1664–5*, pp. 470–71. La Haye, 1893: Nijhoff.

Huygens, C. (1656). *Oeuvres complètes de Christiaan Huygens*, I: *Correspondance 1638–56*, pp. 494–507. La Haye, 1888: Nijhoff.

Huygens, C. (1657). *De ratiociniis in aleae ludo*. In *Exercitationum mathematicarum*, by F. van Schooten. Lugduno Batava: Elsevier. Reprinted (1920) in Huygens' *Oeuvres* (Dutch, with French translation) (see Huygens, 1676), pp. 50–91.

Huygens, C. (1676). *Oeuvres complètes de Christiaan Huygens*, XIV: *Calcul des Probabilités; Travaux de mathématiques pures, 1655–66*, pp. 151–5. La Haye, 1920: Nijhoff.

Kohli, K. (1967). Spieldauer: Von Jakob Bernoullis Lösung der fünften Augfabe von Huygens bis zu den Arbeiten von de Moivre. Universität Zürich dissertation. Printed (1975) in Bernoulli's *Werke* (see Bernoulli, 1684), pp. 403–55.

Montmort, P. (1708). *Essay d'analyse sur les jeux de hazard*. Paris: Quillau.

Montmort, P. (1713). *Essay d'analyse sur les jeux de hazard*, 2nd edn. Paris: Quillau. Facsimile reprint (1980), New York: Chelsea.

Pascal, B. (1665). *Traité du triangle arithmétique*. Paris: Desprez. Reprinted in *Oeuvres complètes* (many different editions).

Thatcher, A. R. (1957). A note on early solutions of the problem of the duration of play. *Biometrika* **44**, 515–18. Reprinted (1970) in *Studies in the History of Statistics and Probability*, Ed. E. S. Pearson and M. G. Kendall, pp. 127–30. London: Griffin.

Todhunter, I. (1865). *A History of the Mathematical Theory of Probability*. Cambridge: Macmillan. Facsimile reprint (1965), New York: Chelsea.

---

* Reprinted in the present volume, Appendix 1.
† This refers to the present book.

REFERENCES

"People ask if I have myself read all the books I quote.—I reply that I have not; it would certainly have meant spending my life reading very bad books; but I read Escobar right through twice; and, as for the others, I got my friends to read them, but I did not use a single passage without reading it myself in the book quoted, going into the context involved, and reading the passage before and after it, to avoid all risk of quoting an objection as an answer, which would have been reprehensible and unjust."

PASCAL
(translated by Krailsheimer, 1966)

The abbreviation *D.S.B.* in a reference stands for the *Dictionary of Scientific Biography*, published in sixteen volumes between 1970 and 1980, C. C. Gillispie, Editor in Chief. New York: Charles Scribner's Sons. See also p. viii.

References

Anbouba, A. (1975). Al-Samawal. *D.S.B.* **12**, 91–5.

Arbuthnott, J. (1712). An argument for Divine Providence, taken from the constant regularity observ'd in the births of both sexes. *Phil. Trans.* **27**, 186–90. Reprinted in Kendall & Plackett (1977), 30–4.

Arbuthnott, J. (1738). *Of the laws of chance, or, a method of calculation of the hazards of game.* 4th edn, revised by John Ham: London.

Bayes, T. (1764). An essay towards solving a problem in the doctrine of chances. *Phil. Trans.* **53**, 370–418. Reprinted in *Biometrika*, **45**, 293–315 (1958), which is itself reprinted in Pearson and Kendall (1970), 131–53.

Bernoulli, James (1713). *Ars conjectandi.* Basilea. Facsimile reprint: Bruxelles: Culture et Civilisation (1968). Reprinted in *Die Werke von Jakob Bernoulli*, III, 107–286, Basel: Birkhäuser (1975).

Bernoulli, John (1742). *Opera omnia*, IV. Lausannae & Genevae.

Bhaskara (1150). *Lilavati*, translated by Colebrooke (1817).

Boethius, A. M. S. (510). *In Isagogen Porphyrii commenta*, ed. G. Schepss and S. Brandt. *Corpus Scriptorum Ecclesiasticorum Latinorum*, **48**, Vienna: Tempsky, and Leipzig: Freytag (1906).

Bosmans, H. (1906). Une note historique sur le triangle arithmétique de Pascal. *Ann. Soc. Scient. Bruxelles*, **31**, 65–72.

Bosmans, H. (1923). Pascal et son traité du triangle arithmétique. *Mathesis*, **37**, 455–64.

Bosmans, H. (1924). Sur l'oeuvre mathématique de Blaise Pascal. Supplément à *Mathesis*, **38**, 1–59.

Boyer, C. B. (1943). Pascal's formula for the sums of powers of the integers. *Scripta math.* **9**, 237–44.

Boyer, C. B. (1950). Cardan and the Pascal Triangle. *Amer. Math. Monthly*, **57**, 387–90.

Boyer, C. B. (1963). Pascal: the man and the mathematician. *Scripta math.* **26**, 283–307.

Boyer, C. B. (1968). *A History of Mathematics.* New York: Wiley.

Braekman, W. L. (1980). Fortune-telling by the casting of dice. *Studia neophilologica*, **52**, 3–29.

Brahmegupta (628). *Arithmetic*, translated by Colebrooke (1817).

Briggs, H. (1624). *Arithmetica logarithmica.* Londini.

Briggs, H. (1633). *Trigonometria Britannica.* Goudae: Petrus Rammasenius.

Buckley, W. (1567). *Arithmetica Memorativa.* Bound with *Dialectica*, by John Seton. Londini: Thomas Marsh (1574).

Bussey, W. H. (1917). The origin of mathematical induction. *Amer. Math. Monthly*, **24**, 199–207.

Buteo, I. (1559). *Logistica.* Lugduni.

162

Cajori, F. (1916). *William Oughtred*. Chicago: Open Court.

Cardano, G. (1539). *Practica Arithmetice, & Mensurandi singularis*. Mediolanum. Reprinted in *Opera Omnia*, IV, Lugduni (1663), as *Practica Arithmeticae*.

Cardano, G. (1550). *De subtilitate*. Norimburgo. Reprinted in *Opera Omnia*, III, Lugduni (1663).

Cardano, G. (1560). *Liber de ludo aleae*. Translated by S. H. Gould in Ore (1953).

Cardano, G. (1570). *Opus novum de proportionibus numerorum*. Basilea.

Chakravarti, G. (1932). Growth and development of permutations and combinations in India. *Bull. Calcutta Math. Soc.* **24**, 79–88.

Chasles, M. (1860). *Les trois livres des Porismes d'Euclide*. Paris: Mallet-Bachelier.

Clagett, M. (1968). *Nicole Oresme and the Medieval Geometry of Qualities and Motions*. Madison: University of Wisconsin Press.

Clavius, C. (1585). *Commentarius in Sphaeram Ioannis de Sacro Bosco*. Romae.

Colebrooke, H. T. (1817). *Algebra with Arithmetic and Mensuration from the Sanscrit of Brahmegupta and Bhascara*. London: Murray.

Collins, J. (1712). *Commercium Epistolicum*. Londini.

Coumet, E. (1972). Mersenne: dénombrements, répertoires, numérotations de permutations. *Math. Sci. Hum.* **10**, 5–37.

David, F. N. (1962). *Games, Gods and Gambling*. London: Griffin.

De Moivre, A. (1698). A method of raising an infinite multinomial to any given power, or extracting any root of the same. *Phil. Trans.* **19**, 619–25.

De Moivre, A. (1712). De mensura sortis. *Phil. Trans.* **27**, 213–64. Translated by B. McClintock in Hald (1984).

De Moivre, A. (1730). *Miscellanea analytica de seriebus et quadraturis*. Londini.

Dickson, L. E. (1919, 1920, 1923). *History of the Theory of Numbers*. Washington: Carnegie Institution. I. *Divisibility and Primality*. II. *Diophantine Analysis*. III. *Quadratic and Higher Forms*. Facsimile reprint: New York: Chelsea (1971).

Diophantus of Alexandria (250). Diophante d'Alexandrie, *Les six livres arithmétiques et le livre des nombres polygones*. Traduites du grec en français par P. Ver Eecke. Bruges: Desclée, de Brouwer (1926).

Edwards, A. W. F. (1978). Commentary on the arguments of Thomas Bayes. *Scand. J. Statist.* **5**, 116–18.

Edwards, A. W. F. (1982a). Sums of powers of integers—a little of the history. *Math. Gaz.* **66**, 22–8.

Edwards, A. W. F. (1982b). Pascal and the Problem of Points. *Int. Statist. Rev.* **50**, 259–66. Reprinted in the present volume, Appendix I.

Edwards, A. W. F. (1983). Pascal's Problem: The "Gambler's Ruin". *Int. Statist. Rev.* **51**, 73–9. Reprinted in the present volume, Appendix II.

Edwards, C. H. (1979). *The Historical Development of the Calculus*. New York: Springer.

Eisenhart, C. & Birnbaum, A. (1967). Anniversaries in 1966–67 of interest to statisticians. Part II: Tercentennials of Arbuthnot and De Moivre. *Amer. Statist.* **21**, 22–9.

Euler, L. (1755). *Institutiones calculi differentialis*. Ticini (1787).

Euler, L. (1775). Demonstratio theorematis neutoniani de evolutione potestatum binomii pro casibus quibus exponentes non sunt numeri integri. *Novi Commentarii Academiae Scientiarum Imperialis Petropolitanae*, **19**, 103–11.

Euler, L. (1784). De mirabilibus proprietatibus unciarum, quae in evolutione binomii ad potestatem quamcunque evecti occurrunt. *Acta Academiae Scientiarum Imperialis Petropolitanae*, 1781 Part I, 74–111.

Faulhaber, J. (1615). *Mysterium Arithmeticum.*

Faulhaber, J. (1631). *Academia algebrae.* Augspurg.

Fermat, P. de (1636). *Oeuvres de Fermat.* ed. P. Tannery & C. Henry. Paris: Gauthier-Villars. I. *Oeuvres mathématiques diverses—observations sur Diophante* (1891). II. *Correspondance* (1894). III. *Traductions par M. Paul Tannery* (1896). IV. *Compléments par M. Charles Henry* (1912). *Supplément* (1922).

Fermat, P. de (1679). *Varia opera mathematica.* Tolosae. See also Fermat (1636).

Galileo, G. (1620). *Sopra le scoperte dei dadi.* Translated by E. H. Thorne in David (1962), 192–5.

Gardner, M. (1968). *Logic Machines, Diagrams and Boolean Algebra.* New York: Dover.

Gillings, R. J. The mathematics of Ancient Egypt. *D.S.B.* **15**, 681–705.

Goldstine, H. H. (1977). *A History of Numerical Analysis from the 16th through the 19th Century.* New York: Springer.

Greenstreet, W. J. (1927). *Isaac Newton 1642–1727.* London: Bell.

Hacking, I. (1971). Jacques Bernoulli's *Art of Conjecturing. Brit. J. Phil. Sci.* **22**, 209–29.

Hacking, I. (1975). *The Emergence of Probability.* Cambridge University Press.

Hald, A. (1984). A. de Moivre: "De Mensura Sortis" or "On the Measurement of Chance". *Int. Statist. Rev.* **52**, 229–62.

Hara, K. (1962). Pascal et l'induction mathématique. *Rev. d'Hist. Sciences,* **15**, 287–302.

Hardy, G. H. & Wright, E. M. (1954). *An Introduction to the Theory of Numbers,* 3rd edn. Oxford: Clarendon Press.

Haussner, R. (1899). *Wahrscheinlichkeitsrechnung (Ars conjectandi von Jakob Bernoulli, 1713). Ostwald's Klassiker Nr. 107.* Leipzig: Engelmann.

Heath, T. L. (1921). *Greek Mathematics.* Oxford: Clarendon Press.

Hérigone, P. (1634). *Cours mathématique,* II. Paris.

Ho Peng-Yoke (1971). Chu Shih-chieh. *D.S.B.* **3**, 265–71.

Ho Peng-Yoke (1976). Yang Hui. *D.S.B.* **14**, 538–46.

Hoe, J. (1977). Les systèmes d'équations polynômes dans le Siyuan Yujian (1303). *Mémoires de l'institut des hautes études chinoises, Collège de France,* VI.

Hofmann, J. E. (1970). Bernoulli, Jakob. *D.S.B.* **2**, 46–51.

Hofmann, J. E. (1974). *Leibniz in Paris.* Cambridge University Press.

Hutton, C. (1785). *Mathematical Tables.* London: Robinson and Baldwin. The Introduction is reprinted, with minor changes, in Hutton (1812), 278–454.

Hutton, C. (1786). Of the binomial theorem. With a demonstration of the truth of it in the general case of fractional exponents. Tract VI of *Tracts Mathematical and Philosophical.* London. Reprinted, with minor changes, in Hutton (1812), 228–44.

Hutton, C. (1812). *Tracts on Mathematical and Philosophical Subjects,* I. London.

Huygens, C. (1655). *Oeuvres complètes de Christiaan Huygens.* XIV. *Calcul des Probabilités; Travaux de mathématiques pures. 1655–1666.* La Haye: Nijhoff (1920).

Huygens, C. (1656). *Oeuvres complètes de Christiaan Huygens.* I. *Correspondance 1638–1656.* La Haye: Nijhoff (1888).

Itard, J. (1948). Sur la date à attribuer à une lettre de Pierre Fermat. *Rev. d'Hist. Sciences,* **2**, 95–8.

Jacobi, C. G. J. (1834). De usu legitimo formulae summatoriae Maclaurinianae. *J. reine u. angew. Math.* **12**, 263–72.

Jacquard, A. (1974). *Les probabilités.* Paris: Presses Universitaires de France.

Joncourt, E. de (1762). *De la nature et des principaux usages de la plus simple espèce de nombres trigonaux.* La Haye: Husson.

Jones, W. (1706). *Synopsis Palmariorum Matheseos or A New Introduction to the Mathematics.* London.

Jordan, C. (1965). *Calcul of Finite Differences,* 3rd edn. New York: Chelsea.

Kästner, A. G. (1799). *Geschichte der Mathematik,* III. Göttingen.

Kendall, M. G. (1956). The beginnings of a probability calculus. *Biometrika,* **43**, 1–14. Reprinted in Pearson & Kendall (1970).

Kendall, M. G. & Buckland, W. R. (1971). *A Dictionary of Statistical Terms.* 3rd edn. Edinburgh: Oliver & Boyd.

Kendall, M. G. & Plackett, R. L. (1977). *Studies in the History of Statistics and Probability,* II. London: Griffin.

Kircherus, A. (1650). *Musurgia Universalis,* II. Romae.

Kircherus, A. (1669). *Ars Magna Sciendi.* Amstelodami.

Knobloch, E. (1973). Die mathematischen Studien von G. W. Leibniz zur Kombinatorik. *Studia Leibnitiana Supplementa,* XI. Wiesbaden: Steiner.

Knobloch, E. (1976). Die mathematischen Studien von G. W. Leibniz zur Kombinatorik. *Studia Leibnitiana Supplementa,* XVI. Wiesbaden: Steiner.

Knobloch, E. (1979). *Musurgia universalis:* unknown combinatorial studies in the age of baroque absolutism. *Hist. Sci.* **17**, 258–75.

Knuth, D. E. (1973). *Fundamental Algorithms. Vol. I of the Art of Computer Programming,* 2nd edn. Reading, Mass.: Addison-Wesley.

Krailsheimer, A. J. (1966). *Pascal: Pensées. Translated by A. J. Krailsheimer.* Harmondsworth, Middlesex: Penguin.

Lam Lay Yong (1970). On the existing fragments of Yang Hui's Hsiang Chieh Suan Fa. *Archive Hist. Exact Sciences,* **6**, 82–8.

Lam, Lay Yong (1980). The Chinese connection between the Pascal Triangle and the solution of numerical equations of any degree. *Historia Mathematica,* **7**, 407–24.

Leibniz, G. W. (1666). *Dissertatio de arte combinatoria.* Published under the title *Ars combinatoria,* Francofurti (1690).

Leibniz, G. W. (1695). *Commercium philosophicum et mathematicum* (Correspondence between Leibniz and John Bernoulli), I: 1694–1699. Lausannae & Genevae. Bousquet (1745).

Levi ben Gerson (1321). *The Work of the Computer.* (Hebrew text and German translation by G. Lange: *Die Praxis des Rechners.*) Frankfurt am Main (1909).

Lohne, J. A. (1965). Thomas Harriot als Mathematiker. *Centaurus,* **11**, 19–45.

Lohne, J. A. (1972). Harriot, Thomas, *D.S.B.* **6**, 124–9.

Lohne, J. A. (1979). A survey of Harriot's scientific writings. *Arch. Hist. Exact Sci.,* **20**, 265–312.

Mahoney, M. S. (1973). *The mathematical career of Pierre de Fermat.* Princeton University Press.

Maistrov, L. E. (1974). *Probability Theory: A Historical Sketch.* New York: Academic Press.

Maseres, F. (1795). *Mr. James Bernoulli's Doctrine of Permutations and Combinations and some other useful mathematical tracts.* London. Also in Maseres (1796), 25–98.

Maseres, F. (1796). *Scriptores logarithmici, Vol. III*. London.

Maurolico, F. (1575). *Arithmeticorum libri duo*. Second part of *Opuscula mathematica*, Venetiis.

Mersenne, M. (1625). *La verité des sciences*. Paris: Toussainct du Bray.

Mersenne, M. (1636). *Harmonicorum libri XII*. Lutetiae Parisiorum.

Mersenne, M. (1637). *Harmonic universelle, Seconde partie*. Paris: Ballard.

Mersenne, M. (1648). *Correspondance du P. Marin Mersenne*, VI, 1636–1637, ed. C. de Waard. Editions de Centre National de la Recherche Scientifique, Paris (1960).

Mesnard, J. (1952). *Pascal, his Life and Works*. Translated by G. S. Fraser. London: Harvill Press.

Miller, J. C. P. (1954). *Table of Binomial Coefficients*. Royal Society Mathematical Tables, III. Cambridge University Press.

Montmort, P. de (1708). *Essay d'analyse sur les jeux de hazard*. Paris: Quillau.

Montmort, P. de (1713). *Essay d'analyse sur les jeux de hazard*, 2nd edn. Paris: Quillau. Facsimile reprint: New York: Chelsea (1980).

Mortimer, E. (1959). *Blaise Pascal, the life and work of a realist*. London: Methuen.

Napier, J. (1594). *De arte logistica*. Edinburgh (1839).

Natucci, A. (1974). Mengoli, Pietro. *D.S.B.* **9**, 303–4.

Needham, J. (1959). *Science and Civilisation in China*, III. *Mathematics and the Sciences of the Heavens and the Earth*. Cambridge University Press.

Newton, I. (1664). *The Mathematical Papers of Isaac Newton*, I, *1664–1666*, ed. D. T. Whiteside. Cambridge University Press (1967).

Newton, I. (1674). *The Mathematical Papers of Isaac Newton*, IV, *1674–1684*, ed. D. T. Whiteside. Cambridge University Press (1968).

Newton, I. (1676). *The correspondence of Isaac Newton*, II, *1676–1687*, ed. H. W. Turnbull. Cambridge University Press (1960).

Nicomachus of Gerasa (100). *Introduction to Arithmetic*. Translated into English by M. L. D'Ooge with Studies in Greek Arithmetic by F. E. Robbins & L. C. Karpinski. New York: Macmillan (1926).

Nunn, T. P. (1910). The arithmetic of infinities. *Math. Gaz.* **5**, 345–56 & 377–86.

Ore, O. (1953). *Cardano, the Gambling Scholar*. Princeton University Press. Reprinted New York: Dover (1965).

Ore, O. (1960). Pascal and the invention of probability theory. *Amer. Math. Monthly*, **67**, 409–19.

Oughtred, W. (1631). *Clavis Mathematicae*. London: Harper.

Oughtred, W. (1652). *Clavis Mathematicae*, 3rd edn. Oxoniae: Lichfield.

Paciolo, L. (1494). *Summa de arithmetica geometria proportioni et proportionalita*. Toscolano (1523).

Pappus of Alexandria (320). Pappus d'Alexandrie, *La collection mathématique*. Traduite du grec un français par P. Ver Eecke. Paris & Bruges: Desclée, de Brouwer (1933).

Pascal, B. (1654a). *Oeuvres complètes*, ed. L. Lafuma. Paris: Seuil (1963).

Pascal, B. (1654b). *Oeuvres complètes*, ed. J. Mesnard. Bruges: Desclée De Brouwer. Vol. I, 1964; II, 1970.

Pascal, B. (1665). *Traité du triangle arithmétique*. Paris: Desprez.

Pearson, E. S. & Kendall, M. G. (1970). *Studies in the History of Statistics and Probability*. London: Griffin.

Porphyry (275). *Isagoge*. Traduction par J. Tricot. Paris: Vrin (1947).

Prestet, J. (1675). *Elemens des Mathematiques*. Paris: Pralard.

Rabinovitch, N. L. (1970a). Studies in the History of Probability and Statistics. XXIV. Combinations and probability in rabbinic literature. *Biometrika*, **57**, 203–5. Reprinted in Kendall & Plackett (1977).

Rabinovitch, N. L. (1970b). Rabbi Levi ben Gershon and the origins of mathematical induction. *Arch. Hist. Exact Sci.* **6**, 237–48.

Rabinovitch, N. L. (1973). *Probability and Statistical Inference in Ancient and Medieval Jewish Literature*. University of Toronto Press.

Rashed, R. (1973). Al-Karaji. *D.S.B.* **7**, 240–6.

Rouse Ball, W. W. (1908). *A Short Account of the History of Mathematics*. 4th edn reprinted, New York: Dover (1960).

Runes, D. D. (1962). *Classics in Logic*. New York: Philosophical Library.

Sarton, G. (1947). *Introduction to the History of Science*, III: *Science and Learning in the Fourteenth Century*, Part I. Baltimore: Williams and Wilkins.

Sarton, G. (1958). *Six Wings*. London: Bodley Head.

Scheubelius, J. (1545). *De numeris et diversis rationibus*. Lipsiae.

Schneider, I. (1982). Jakob Bernoulli und Johannes Faulhaber über arithmetische Reihen höherer Ordnung. *Jahrbuch 1982 der Technischen Universität München*, 132–40.

Schneider, I. (1983). Potenzsummenformeln im 17. Jahrhundert. *Hist. Mathematica*, **10**, 286–96.

Schwenter, D. (1636). *Deliciae Physico-Mathematicae*. Nürnberg.

Scott, J. F. (1938). *The Mathematical Work of John Wallis*. London: Taylor and Francis. Reprinted New York: Chelsea (1981).

Shafer, G. (1978). Non-additive probabilities in the work of Bernoulli and Lambert. *Arch. Hist. Exact Sci.*, **19**, 309–70.

Sheynin, O. B. (1977). Early history of the theory of probability. *Arch. Hist. Exact Sci.*, **17**, 201–59.

Singh, A. N. (1936). On the use of series in Hindu mathematics. *Osiris*, **1**, 606–28.

Smith, D. E. (1923). *History of Mathematics*, I: *General Survey of the History of Elementary Mathematics*. Reprinted New York: Dover (1958).

Smith, D. E. (1925). *History of Mathematics*, II: *Special Topics of Elementary Mathematics*. Reprinted New York: Dover (1958).

Smith, D. E. (1929). *A Source Book in Mathematics*. New York and London: McGraw-Hill.

Spirto, L. (1535). *Libro della Ventura*. Romae: Antonio Blado.

Srinivasiengar, C. N. (1967). *The History of Ancient Indian Mathematics*. Calcutta: World Press.

Stifel, M. (1544). *Arithmetica Integra*. Norimburgae.

Stifel, M. (1545). *Deutsche Arithmetica*. Nürnberg.

Strode, T. (1678). *A short treatise of the combinations, elections, permutations & composition of quantities*. London.

Struik, D. J. (1969). *A Source Book in Mathematics, 1200–1800*. Cambridge, Mass.: Harvard University Press.

Tartaglia, N. (1556). *General trattato di numeri, et misure*. Vinegia.

Taton, R. (1974). Pascal, Blaise, *D.S.B.* **10**, 330–42.

Todhunter, I. (1865). *A History of the Mathematical Theory of Probability*. Cambridge: Macmillan. Reprinted New York: Chelsea (1965).

Tropfke, J. (1924). *Geschichte der Elementar-Mathematik*. VI: *Analysis, Analytische Geometrie*. Berlin & Leipzig: De Gruyter.

Turnbull, H. W. (1933). James Gregory: a study in the early history of interpolation. *Proc. Edinb. Math. Soc. Ser. 2*, **3**, 151–72.

Van der Waerden, B. L. (1981). Mathematics and astronomy in Mesopotamia. *D.S.B.* **15**, 667–80.

Van Schooten, F. (1657). *Exercitationum mathematicarum*. Lugduno Batava: ex Johannis Elsevirii.

Vieta, F. (1591). *Opera Mathematica*, ed. F. van Schooten, Lugduni Batavorum: Elzeviriorum. For a translation see *The Analytic Art*, trans. T. R. Witmer, Kent State University Press, 1983.

Wallis, J. (1655). *Arithmetica infinitorum*. In *Opera mathematica*, I, 355–478, Oxoniae (1695).

Wallis, J. (1685a). *A Treatise of Algebra*. London.

Wallis, J. (1685b). *A Discourse of Combinations, Alternations, and Aliquot Parts*. London. Bound with Wallis (1685a).

Wang, L. & Needham, J. (1955). Horner's method in Chinese mathematics: its origins in the root-extraction procedures of the Han dynasty. *T'oung Pao*, **43**, 345–401.

Waxman, M. (1960). *A History of Jewish Literature*, I. New York and London: Yoseloff.

Whiteside, D. T. (1961a). Patterns of mathematical thought in the later seventeenth century. *Arch. Hist. Exact Sci.*, **1**, 179–388.

Whiteside, D. T. (1961b). Newton's discovery of the general binomial theorem. *Math. Gaz.* **45**, 175–80.

Whiteside, D. T. (1967), *see* Newton (1664).

Whiteside, D. T. (1971), *see* Newton (1674).

Whiteside, D. T. (1972). Gregory, James. *D.S.B.* **5**, 524–30.

Yadegari, M. (1980). The binomial theorem: a widespread concept in medieval Islamic mathematics. *Hist. Mathematica*, **7**, 401–6.

Youschkevitch, A. P. (1971). Euler, Leonhard. *D.S.B.* **4**, 467–84.

Youschkevitch, A. P. & Rosenfeld, B. A. (1973). Al-Kashi, *D.S.B.* **7**, 255–62.

INDEXES

The notes at the end of each chapter, and the references, are not indexed

General Index

170

Index of names

75/14

67/04